# Geography of the Physical Environment

The *Geography of the Physical Environment* book series provides a platform for scientific contributions in the field of Physical Geography and its subdisciplines. It publishes a broad portfolio of scientific books covering case studies, theoretical and applied approaches as well as novel developments and techniques in the field. The scope is not limited to a certain spatial scale and can cover local and regional to continental and global facets. Books with strong regional focus should be well illustrated including significant maps and meaningful figures to be potentially used as field guides and standard references for the respective area.

The series appeals to scientists and students in the field of geography as well as regional scientists, landscape planners, policy makers, and everyone interested in wide-ranging aspects of modern Physical Geography. Peer-reviewed research monographs, edited volumes, advance and undergraduate level textbooks, and conference proceedings covering the major topics in Physical Geography are included in the series. Submissions to the Book Series are also invited on the theme 'The Physical Geography of…', with a relevant subtitle of the author's/editor's choice. Please contact the Publisher for further information and to receive a Book Proposal Form.

Roland Baumhauer

# Physical Geography of Africa

Springer

Roland Baumhauer
Institut für Geographie und Geologie
University of Würzburg
Würzburg, Germany

ISSN 2366-8865 ISSN 2366-8873 (electronic)
Geography of the Physical Environment
ISBN 978-3-662-71405-8 ISBN 978-3-662-71406-5 (eBook)
https://doi.org/10.1007/978-3-662-71406-5

This book is a translation of the original German edition "Die Physische Geographie Afrikas" by Roland Baumhauer, published by Springer-Verlag GmbH, DE in 2023. The translation was done with the help of an artificial intelligence machine translation tool. A subsequent human revision was done primarily in terms of content, so that the book will read stylistically differently from a conventional translation. Springer Nature works continuously to further the development of tools for the production of books and on the related technologies to support the authors.



Planung/Lektorat: Simon Shah-Rohlfs

This Springer imprint is published by the registered company Springer-Verlag GmbH, DE, part of Springer Nature.
The registered company address is: Heidelberger Platz 3, 14197 Berlin, Germany

# Preface

The idea to compile this book resulted from my experience that there are virtually no up-to-date overviews of physical geography of Africa. Africa, with an area of over 30 million square kilometers, the second largest continent, accounting for about 20% of the total land area, offers unparalleled insights into the oldest geological strata, is home to the largest desert, and is enormously rich in mineral resources. These are some of the few things we have all learned and know about the physical geography of Africa. But it is worth taking a deeper look at the continent, because Africa is closer to us and much more important than we would like to admit. Against this background, an overall presentation of the physical geography of the continent seems overdue.

This book is intended primarily for anybody who is interested in basic physical geography of Africa. The aim is to provide an overall look at a relatively durable statement of physical conditions in Africa. The volume includes a presentation of the main environmental factors (geology, geomorphology, climate, soil types, etc.), a discussion of the main types of environment (deserts, savannas, lakes etc.) with an emphasis on some of the linkages between different components of the environment, and a consideration of the main environmental issues related to human use of the land (e.g. soil erosion, desertification).

The project of writing such a book cannot be achieved without the help of numerous people. I would like to thank all the people who played a role in the production of this book. A special thank you goes to the highly appreciated cartographer Julia Breunig. Finally, I would like to thank my family and particularly my spouse for her support of my efforts and her tolerance of my preoccupations.

Würzburg, Germany Roland Baumhauer

**Competing Interests** The author has no competing interests to declare that are relevant to the content of this manuscript.

# Contents

# About the Author

**Roland Baumhauer** was Professor of Geography at the Julius-Maximilians University of Würzburg. Since 2022, he has been Senior Professor of Physical Geography. He has worked extensively in Western and North Africa, and his main research interests are in the fields of geomorphology and environmental change, especially in the Old World dry belt.

# The Natural Environments of the Continent

# 1

Africa, with an area of over 30 million square kilometers is the second largest continent, accounting for about 20 per cent of the total land area (for area comparison see Fig. 1.1). The continent extends in a north–south direction between Cape Blanc in Tunisia (37° 51′ N) and Cape Agulhas in the Republic of South Africa (34° 51′ S) over about 8,000 km almost equidistant from the equator. With approximately 7,500 km from Cap Vert in Senegal to Cape Guardafui in Somalia, Africa's maximum west–east extension nearly reaches the north–south extent. At the equator, the land mass extends about 3,500 km from west to east, and at the southern tropic over 2,100 km. With respect to latitude, the larger part of the continent lies in the Northern Hemisphere.

Separated by the South Atlantic, its western neighboring continent is South America, which is 3,000 km away. Antarctica is about 4,000 km from the southern tip of Africa (from the southern tip of South America, it is only about 1,200 km to Antarctica). In the north, the Mediterranean separates Africa from Europe. At the narrowest point, the Strait of Gibraltar, the Iberian Peninsula is only 14 km away. The Red Sea and the Sinai Peninsula, as well as the Indian Ocean border the mainland of the continent to the east.

Noteworthy is Africa's scarcely structured coastline, which at 36,000 km is about as long as the coast of Europe, which is three times smaller. Although the coastal forms are by no means uniformly structured, Africa has the clearest outline of all continents.

Like no other continent, Africa offers insights into the oldest geological history of the Earth. Almost 60 per cent of the continental area consists of the directly exposed **Precambrian basement complex**, which is older than 550 million (Ma) years. But also, the geologic set-up of the remaining 40 per cent of Africa's surface is formed by Precambrian basement under younger coverage. Only the extreme southwest and the extreme northwest of Africa show Phanerozoic fold belts. The Mediterranean portion of the **Atlas Mountains** in North Africa, began forming approximately 150 million years ago during the late Mesozoic era, resulting from the collision between the African and Eurasian tectonic plates. In contrast, the folding of the **Cape Ranges of South Africa** occurred between 280 and 230 million years ago, along the southwestern fringe of the ancient supercontinent **Gondwana**. Among the geologically youngest formations in Africa are the volcanic and sedimentary deposits, which formed as a result of rifting processes in the **Central- and East African Rift Valley**. The Palaeogene and Neogene basin sediments in regions like the Murzuk, Congo, and Kalahari Basins are also relatively young, dating back to these epochs.

Within the African continent itself, the first-order morphological distinction is between the broadly elevated south and east, and the



R. Baumhauer, *Physical Geography of Africa*, Geography of the Physical Environment,
https://doi.org/10.1007/978-3-662-71406-5_1

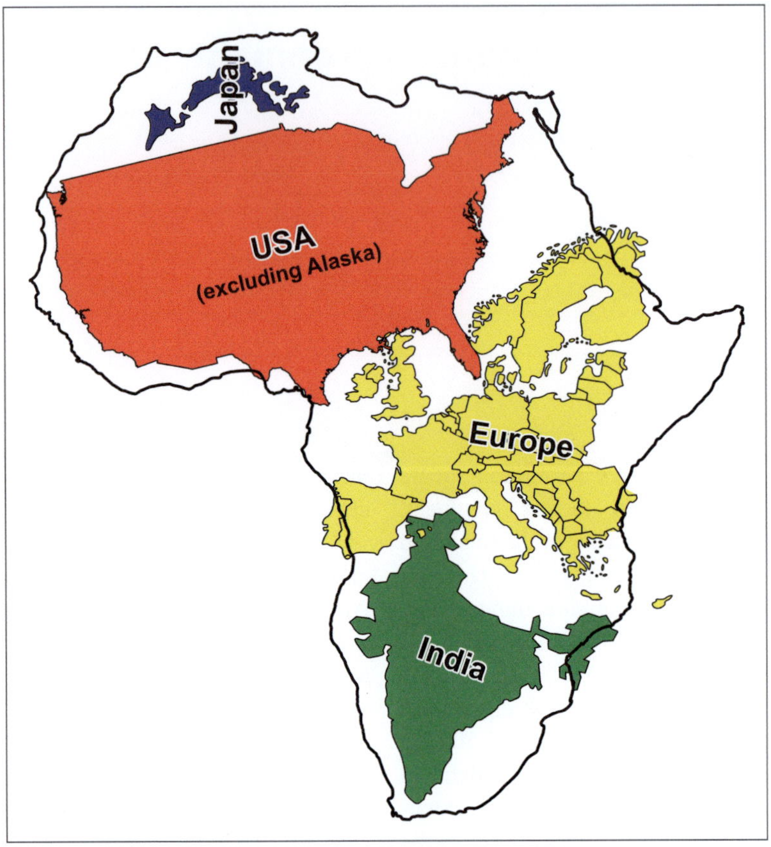


**Fig. 1.1** Size of Africa compared to the USA (excluding Alaska), Japan, India, and Europe (EU including Norway, United Kingdom, Switzerland, and the Balkan countries)

lower-lying north and west. A detailed assessment of the height-frequency distribution of Africa reveals a concentration of elevation between 400 and 600 m, with a secondary peak in the distribution between 800 and 1,000 m. Extensive lowlands as well as alpine mountains with elevations over 4,000 m are missing. The significance of such height-area distribution characteristics lies in the way they reflect the tectonic and long-term landscape evolution of the continent. According to the land surface altitude, the continent can be divided into a low African and a high African part.

**Low Africa** includes North Africa south of the Atlas Mountains as well as West and Central Africa (Fig. 1.2). Its mean elevation with respect to sea level is 300 m, and it is divided into geological swells (or ridges) and basins. The Atlas chains of Northwest Africa form the structural continuation of the European folded chains, such as the Betic Cordillera on the Iberian Peninsula, the Pyrenees or the Alps, which have emerged since the end of the Mesozoic during the alpine building. **High Africa** begins in the Highlands of Ethiopia and continues over the highlands of East Africa and the small mountain massif of Lesotho to the southern tip of the continent (Fig. 1.2). Large parts of Eastern and Southern Africa lie above 1,000 m. The mean elevation of this part of Africa is 1,200 m and reaches the highland massifs at a mean height of 2,500 m. Just like in Low Africa, ridges and basins are also the first-order morphological characteristics in High Africa, for example, the Kalahari Basin or the Lunda Ridge. As a result of its elevated position and unique genetic characteristics, the Rift Valley, a vast, 6,500 km-long tectonic trench system, exhibits

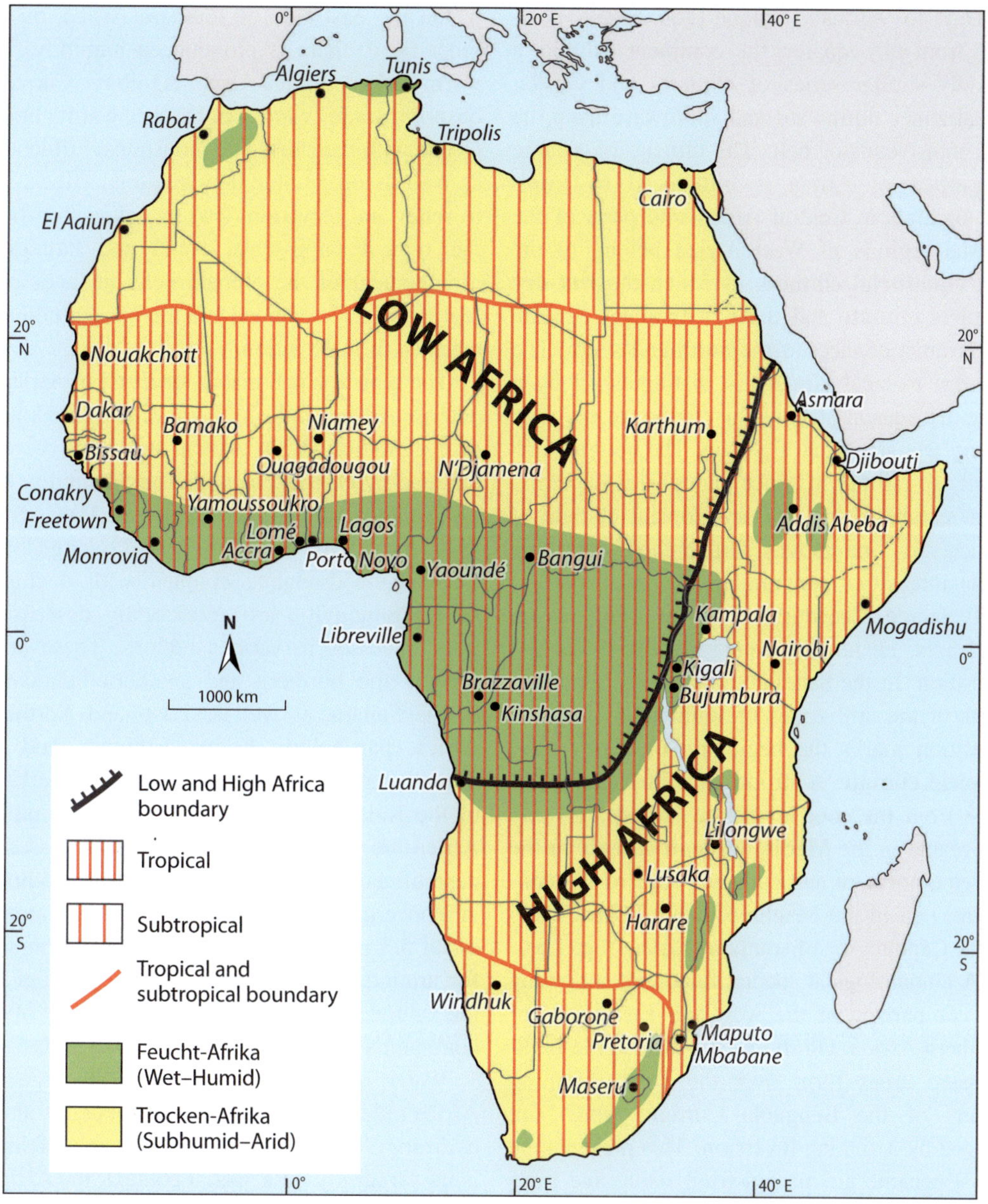


**Fig. 1.2** The main physiographic structures of Africa. The mean altitude of Lower Africa is 300 m a.s.l., that of High Africa 1,200 m a.s.l. The distribution of land—surface altitude: <200 m a.s.l.: 10%; 200–1,000 m a.s.l.: 63.5%; 1,000–2,000 m a.s.l.: 23%; >2,000 m a.s.l.: 3.5% of the total area

a distinctly differentiated structure. Specifically, the southern Rift, and the eastern African highlands, comprising the Central African and East African Rift, exhibit a more pronounced and finely differentiated geological structure. This rift system continues northward into the Red Sea and further over the Dead Sea to the Lebanon-Anti-Lebanon mountains. The volcanic massifs associated with the rift system, such as Kilimanjaro (5,895 m), Mount Kenya (5,199 m), Virunga (4,507 m), or the Ruwenzori (5,119 m), as well as the vast flat basin of Lake Victoria, the third largest lake in the world by water surface, contribute to this.

Due to Africa's unique equidistant position from the equator, the continent exhibits a broadly similar series of climates and vegetational zones northward and southward from the hot, moist equator belt. The climate of a large proportion of Africa is influenced by tropical circulation. Central Africa and parts of the coastal regions of West Africa belong to the **wet equatorial climate**, to which the **wet-dry tropical climate** and the **dry tropical climate** of Savanna connect to the north and south. The wet-dry tropical climate is distinguished by a very dry season at low sun that alternates with a very wet season at high sun. The natural variability in rainfall in space and time is a permanent feature of the wet-dry tropical climate of Africa. As one moves farther poleward, this cycle intensifies and rainfall becomes increasingly scarce. The duration of the rainy season decreases, eventually giving way to a distinct transition. In the Sahelian-Saharan region to the north or the arid zones of southern Africa, this transition marks the beginning of a **dry subtropical climate**. A stronger extratropical influence from the cool temperate latitudes is only noticeable in the **Mediterranean climate** at the extreme northern and southern edge of the continent, i.e., in the Maghreb of north Africa and in the Cape region of southern Africa (Fig. 1.3).

A climatological special feature is shown by the comparison of the west and east coasts of southern Africa. On the west coast, stable high-pressure areas form over the cold upwelling waters of the Benguela Current, which are capped by a sinking inversion. This prevents the moist oceanic air masses from rising and raining down, resulting in dense coastal fog. This effect is further enhanced by the offshore trade winds. Large-scale arid regions are created, in places even pronounced coastal deserts, such as the Namib in Namibia. The same phenomenon is also evident along the west coast of northern Africa, however, the coastal desert characteristics, shaped by the cool Canary Current and dry offshore trade winds, are largely overshadowed by the dominant macroclimatic conditions of the Sahara Desert, which stems from the vast adjacent landmass.

On the east side of southern Africa, on the other hand, there is pronounced humidity. The warm Mozambique and Agulhas Currents, onshore trade winds from the warm Indian Ocean with their seasonal monsoonal influences, as well as orographic and frictional effects due to relief and elevation, lead to a deviation from the typical large-scale, zonal and latitudinal arrangement of the climate belts in favor of a regionally differentiated and more longitudinally oriented differentiation.

The typical zonal macro-structure of Africa is determined not only by height-area distribution characteristics (High Africa and Low Africa), but also by climatic factors. Thus, humidity and aridity allow a distinction between **Wet Africa** and **Dry Africa** (see Fig. 1.2). Prolonged and intermittent droughts, coupled with declining land productivity, are accelerating desertification processes in various regions. These areas include the northern and southern boundaries of the Sahara, as well as East- and Northeast Africa, particularly Kenya, Ethiopia, and the Somali Peninsula. Additionally, the borderlands of the Kalahari and Namib deserts in southern Africa are also susceptible to the processes of desertification, posing significant environmental concerns. These regions comprise an area of about 5.3 million square kilometers, accounting for around 30 per cent of Africa's area outside the natural deserts, with about 65 per cent of the continent's total population.

Water is the decisive natural resource for Africa's development. Looking generally at the hydrology of Africa, its most distinctive feature is its variability. In a spatial context, the African continent arguably embraces greater contrast than any other continent. About 43 per cent of Africa's population south of the Sahara does not have a sufficient water supply. The runoff behavior and runoff volume of the rivers are closely linked to the climate and the morphology of the African continent. With the wide range of climates existing over the continent and the associated variability in the seasonal distribution of precipitation (and evapotranspiration), African rivers exhibit a wide range of flow patterns. The wet-equatorial climate of Central and

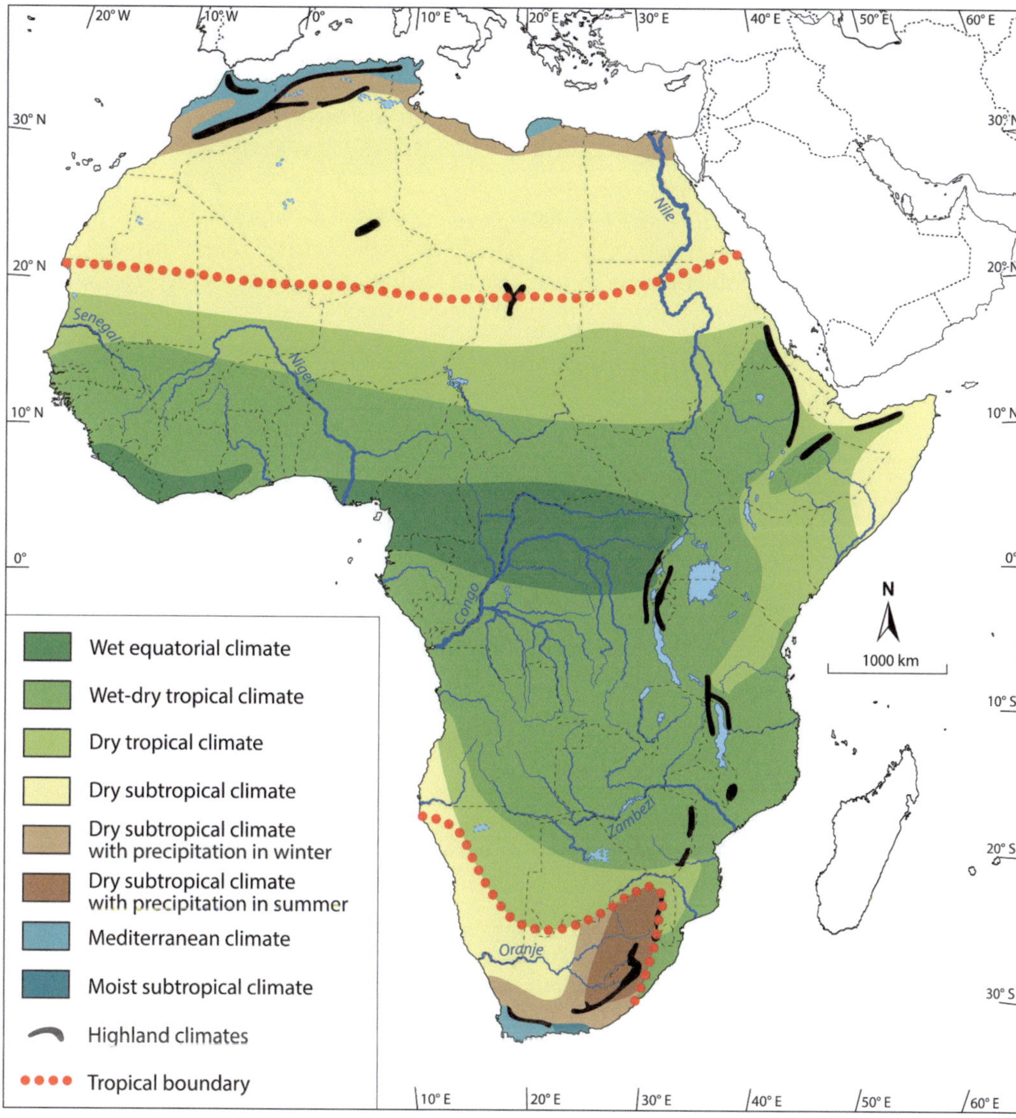


**Fig. 1.3** Climates of Africa

West Africa has abundant water all months, and the traditional flow regime show perennial rivers like Congo, Ubangi, or the lower Niger. In the wet-dry tropical climate, the flow pattern is dominated by the annual high water during the rainy season in late summer and a period of low flows or a complete dry fall in the dry low-sun season in winter. These rainy season runoff maxima can also shift to the beginning of the dry season, for example, at the lower Zambezi River or Okavango River.

The wetlands of Africa are concentrated south of the Sahara and north of the Tropic of Capricorn and the Kalahari, in the boundary of Lake Chad, on the Bahr al-Ghazal in South Sudan, on the Okavango in southern Africa or the Kafue Flats in Zambia. They have water tables at or near the surface, either seasonally or permanently. On the African continent, which is semi-arid or arid over much of its area, wetlands, especially in Lower Africa have long been of great importance to its people and wildlife.

Only episodic runoff occurs in the dry valleys of the desert areas of the continent, in the wadis (or Ennedis of the Sahara, or the Rivieres, Omiramba or Lishana of southern Africa). While these valleys often only flow through once in several years, the subtropical winter rain areas show the typical torrential runoff regime of the Mediterranean region. In the Maghreb, a further smaller peak in runoff often occurs in late spring, caused by the snowmelt in the Atlas Mountains. For the water balance of East Africa, the system of Lake Victoria with the connected lakes and the wetland areas influenced by the White Nile in Uganda up to the South Sudanese "Sudd", one of the largest wetlands on earth, is of hydrological importance. Due to the latitude-parallel climate zones and the macroscale morphology features, Africa offers the

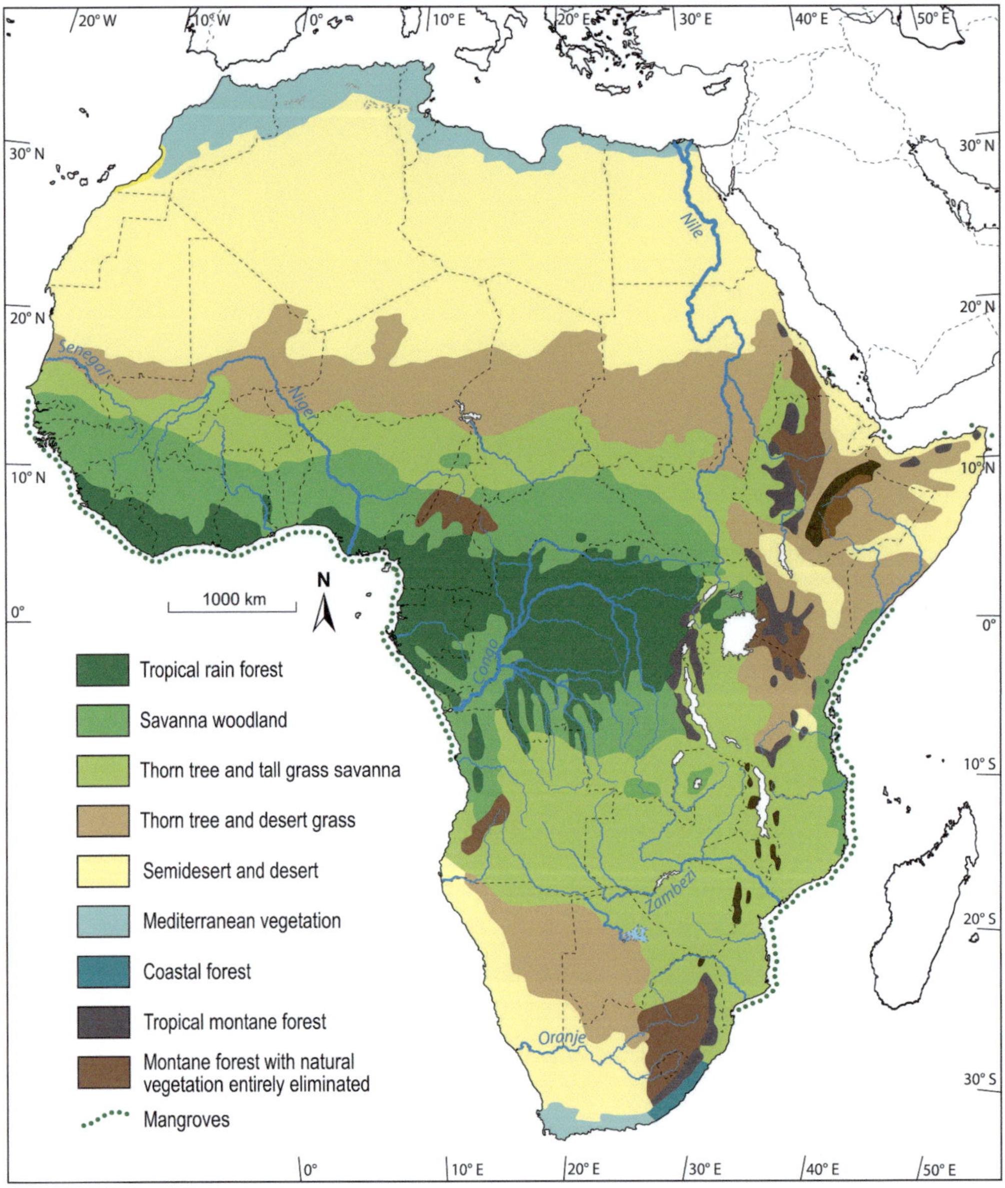


**Fig. 1.4** The natural vegetation of Africa

opportunity to consider the vegetational zones of the tropics and the adjoining subtropics in a model-like manner (see Fig. 1.4). The biomes are still important to this day for understanding the different ways of life and economic forms in Africa. Less important than in the other two tropical continents, Asia and South America is the mountain biota of Africa. It is only found in the highland massifs of Ethiopia and Lesotho up to the Afro-alpine as well as in the volcanic massifs of East and West Africa up to the naval altitude level.

## Further Readings


Balek J (1977) Hydrology and water resources in tropical Africa. Elsevier, Amsterdam, p 435

Bloom DE, Sachs JD, Collier P, Udry C (1998) Geography, demography, and economic growth in Africa. Brook Pap Econ Act 2:207–295

Boateng EA (1980) A political geography of Africa. Cambridge University Press, Cambridge, p 320

Bridges EM (1990) World geomorphology. Cambridge University Press, Cambridge, p 260

Collier P (2006) Africa: geography and growth. Centre for the Study of African Economies, Department of Economies, Oxford University, p 12

Goudie AS (1996) Climate: past and present. In: Adams WM, Goudie AS, Orme AR (eds) The physical geography of Africa. Oxford University Press, New York, pp 35–59

Luig U, Oppen AV (1997) The making of African landscapes. Paideuma 43:7–252

Manshard W (1981) Afrika—südlich der Sahara. Fischer, Frankfurt, p 328

Meadows ME (1996) Biogeography. In: Adams WM, Goudie AS, Orme AR (eds) The physical geography of Africa. Oxford University Press, New York, pp 161–172

Mensching H, Wirth E (1989) Nordafrika und Vorderasien. Fischer Taschenbuch Verlag, Frankfurt, p 317

Mountjoy AB, Hilling D (1987) Africa: geography and development. Taylor & Francis, Routledge Revivals, London, p 465

Nash DJ, Meadows ME (2012) Africa. In: Metcalfe SE, Nash DJ (eds) Quaternary environmental change in the tropics. John Wiley & Sons, Hoboken, pp 79–150

Nunn N, Puga D (2009) Ruggedness: the blessing of bad geography in Africa. National Bureau of Economic Research, Cambridge, p 34

Stock R (2013) Africa south of the Sahara: a geographical interpretation, 3rd edn. The Guilford Press, New York, p 593

Tolmacheva M (2008) Environment and nature in Africa. In: Selin H (eds) Encyclopaedia of the history of science, technology, and medicine in non-western cultures. Springer, Dordrecht, pp 751–757

United Nations Environment Program (2008) Africa: atlas of our changing environment. UNEP, Nairobi, p 375

Wiese B (1985) Südafrika. In: Klimm E (ed) Africa 2. List, Munich, pp 233–258

Wiese B (1997) Afrika: Ressourcen, Wirtschaft, Entwicklung. Teubner, Stuttgart, p 262

# 2 The Major Regions of the Continent

After a general overview of Africa in chapter one, a first physisogeographic approach of major regions of the continent is now considered: North Africa, West and Central Africa, East Africa, Zambezi countries, and South Africa (see Fig. 2.1).

**North Africa** is used differently as a geographical term. This greater area includes the Maghreb states Morocco, Algeria, and Tunisia in the west (Maghreb = derived from Arabic west), which are dominated by the approximately SW-NE-striking, maximum 250 km wide Atlas Mountains, and the eastward to the Nile Delta following the Mediterranean Sea coast and its hinterland in Libya and Egypt. A total area of around six million square kilometers. The northern boundary of the **Sahara** lies on the southern piedmont of the Atlas Mountains and the hinterland of the Mediterranean Sea coast. The Sahara is the world's largest desert region and covers an area of approximately seven million square kilometers, as large as the continental United States and Alaska combined. It covers about a quarter of the African continent and extends from the Atlantic coast (at 17° W) to the Red Sea (37° E) over 5,500 km and from the Mediterranean to the Niger River bend and Lake Chad and the margins of the Ethiopian Highlands over 1,500 km. The desert area is divided among eleven states. The largest part is taken by the Mediterranean neighboring countries Algeria and Libya as well as Mauritania and Mali, whose borders reach north beyond the Tropic of Cancer. All Sahara countries have their main settlement regions outside the desert. For the North African countries, these are mainly the Mediterranean, but also the Atlantic coastal areas. The sub-Saharan Sahel is oriented towards the south to the Gulf of Guinea. The Sahara itself represents a physiogeographic, economic, and social border area.

The Sahara's climate is characterized by arid to hyper-arid, with vast expanses of the central and eastern parts of the desert receiving less than 100 mm of rainfall annually. Traditionally, the northern boundary can be drawn with the 100 mm-isoline of annual precipitation (isohyet). The southern boundary of the Sahara is set at the 150 mm annual isohyet. The higher annual rainfall at the southern edge of the Sahara, results from the rainy season in the high-sun period in summer, while the northern boundary of the desert receives its rain in the low-sun period in winter.

The Sahara is characterized by sparse and monotonous vegetation, with many areas, e.g. the Hammadas (vast, rocky plateaus where the sand has been eroded over millennia, leaving behind only stones and pebbles) and the sand sheets of the central Sahara of Niger, being almost completely devoid of vegetation. From north to south, there are transitions from the Mediterranean vegetation of the Maghreb region to tropical vegetation types to the south and the east. Characteristic is the



R. Baumhauer, *Physical Geography of Africa*, Geography of the Physical Environment,
https://doi.org/10.1007/978-3-662-71406-5_2

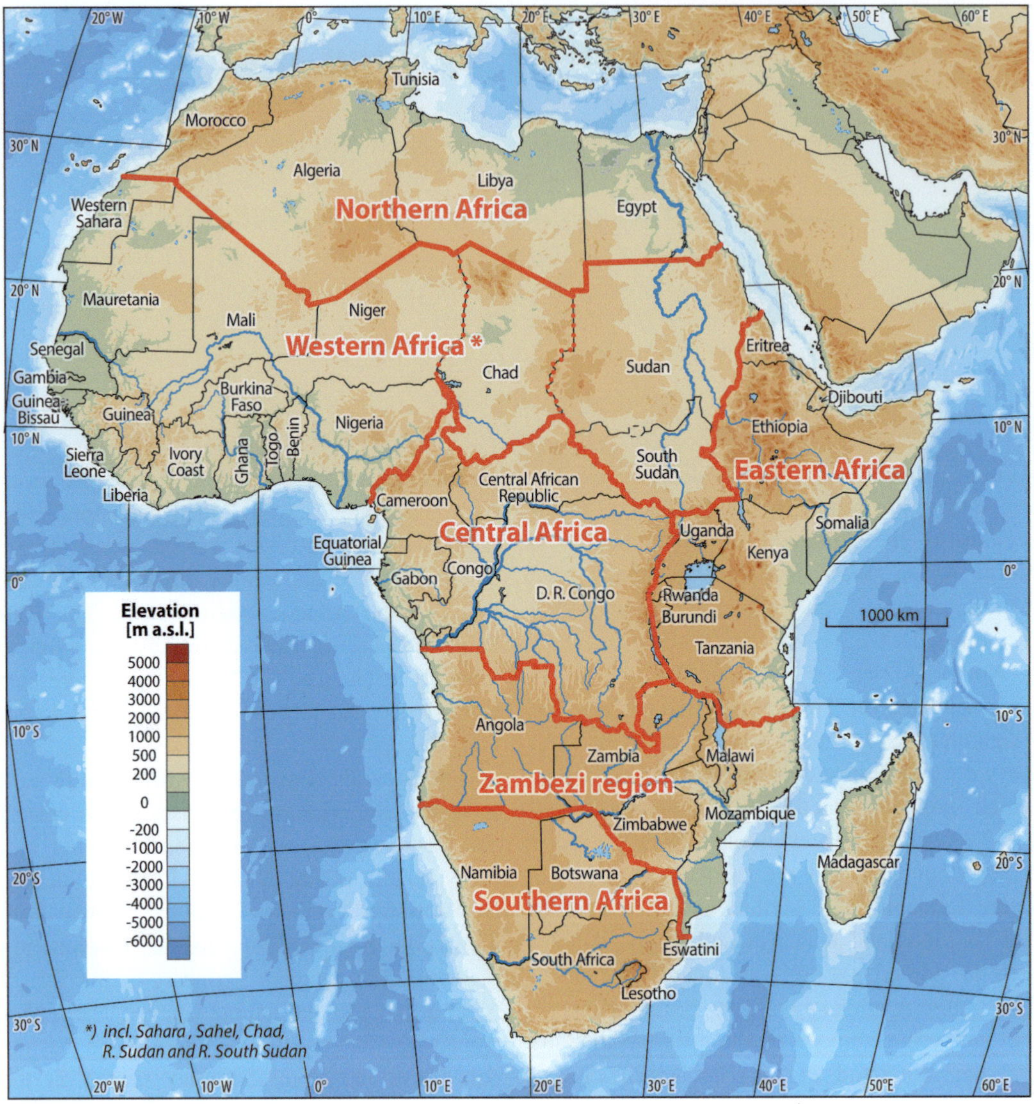


**Fig. 2.1** The major regions of Africa

presence of the productive date palm (*Phoenix dactylifera*), whose distribution in the north is limited by the low winter temperatures in the Atlas Mountains, the Meseta highlands, and in the south by increasing humidity. On the southern margins of the Sahara the Saharan Had shrub (*Cornulaca monacantha*) grads into the Sahelian Cram-Cram grass (*Cenchrus biflorus*), whose floristic analogue at the northern margin of the desert is the Halfa grass (Macrochloa, formerly *Stipa tenacissima*). However, there are broad zones at the margin of the desert that are described as semi-arid semi-desert. This makes a sharp boundary drawing impossible. Rather, the boundary lines differ depending on criteria, type, and processor by up to 200 km.

The Sahara is also one of the hottest regions of the world with average maximum temperatures exceeding 35° covering wide areas, and over 40 °C in the southern part of the region.

The Sahara is characterized by generally moderate relief. The only alpine mountains occur in the central parts of the desert and include the Hoggar, Tibesti, Air, and Ennedi Mountains, which reach elevations of as much as 3,300 m.

South of the Sahara are the two major regions **West- and Central Africa**. They extend from the Atlantic Ocean in the west to the Red Sea in the east. West Africa includes the coastal regions on the Atlantic in the west and south and stretches east to Mount Cameroon and the Adamawa Highlands in northern Cameroon northeast of the "knee of Africa". Physically geographically, West Africa can be divided into two major areas or zones: **Sahel** and **Sudan (zone)** with the **Guinea coast**.

The **Sahel** (Sahel is etymologically derived from Arabic as-sahil = "the shore, the coast") is the dry northern part of West and Central Africa. The semi-arid Sahel zone grades into the Sahara around 15° N. The approximately 300–500 km wide Sahelian transition zone between the Sahara in the north and the Sudan zone in the south extends from the Atlantic Ocean coast across the African continent to the Red Sea and also includes, concerning physisogeographic features, Djibouti, Somalia and northern Kenya (which, however, not bordering the Sahara, are here treated as a part of East Africa). The Sahel is subject to years of devasting drought as well as years of abnormally high rainfall that can result in severe floods. Climate records show that years of abnormally low rainfall typically alternate with several successive years of average or higher-than-average rainfall. This variability, not only in time but also in space is a permanent feature, resulting in devastating drought disasters with increased wind erosion on the one hand, but strong water erosion in the wet years on the other. Characteristic plants of the Sahel are acacias, such as the Seyal acacia (*Vachellia seyal*), the Egyptian Thorn (*Acacia nilotica*), the Arabian Gum Acacia (*Vachellia nilotica*), but also the "infamous Sahel burr Cram-Cram" (*Cenchrus biflorus*).

Due to these physiogeographic factors and the increasing human pressure on natural resources for survival, the Sahel region is certainly the region on Earth where drought, desiccation and (dry-land) degradation and the resulting desertification are most intensely perceived by the public—not least due to the strong media presence during recurring drought periods and the famines caused by them since the 70s of the last century.

Apart from Mauritania, all Sahel countries also share in the better-watered and denser vegetation-covered **Sudan (zone)** (Sudan is derived etymologically from Arabic Bilad as Sudan = "Land of the Blacks"). From west to east, these are the current states of Senegal, Mali, Burkina Faso, and Niger. Gambia, which is surrounded by Senegal, is also counted as part of this geographical region, as are Chad and Sudan. The physiogeographic region of Sudan also includes the coastal countries of West Africa, which are separated to the east by Mount Cameroon and the natural border of the Adamawa Highlands of Central Africa. British and Portuguese sailors referred to the coastal regions of Africa south of the Sahara to Angola as **Guinea lands**: Upper Guinea for the coastal areas west of Mount Cameroon, Lower Guinea for the coastal regions south of it. While the term "Guinea lands" for Guinea-Bissau, Guinea (Conakry), Sierra Leone, Liberia, Ivory Coast (Republic of Côte d'Ivoire), Ghana, Togo, Benin, and Nigeria has remained to this day, "Lower Guinea" is no longer in use.

The Sudan zone belongs to the wet-dry tropical climate, which is often also referred to as the "outer tropics". Its climatic features are, in addition to the small annual range in temperature, typical for the tropics, the alternation of a wet-sun season with a dry low-sun season and a marked temperature cycle, with the hottest temperatures before the rainy season. The coastal areas of Guinea lands have up to 200 km inland, reaching tropical rainforest areas, and are part of the wet-equatorial climate or the "inner tropics". The northern boundary of the Guinea lands lies on the Atlantic west coast up to 8° N. They stretch along the Gulf of Guinea of more than 3,500 km and show comparable ecological features due to the climate, vegetation, and land use zones parallel to the latitude circle. More obvious is the landscape transformation in the meridional direction from the Guinea coast to the Sahel. The tropical rain forest on the Guinea coast is interrupted by the Dahomey Corridor (also "Dahomey Gap" in Ghana and Togo), a savanna corridor that extends all the way to the coast in Benin, Togo, and Ghana (see Chap. 5).

The geological **Congo Basin** and its marginal ridges (or swells) form the greater area of Africa, which is also referred to as **Central Africa** (or Middle Africa). It includes the states Equatorial Guinea, Gabon, Cameroon, the Democratic Republic of Congo (DR Congo, formerly Zaire), the Republic of Congo (Brazzaville), the Central African Republic as well as the islands of Sao Tomé and Principe. They all have shares in the wet tropics and are still among the most thinly populated sub-Saharan regions of Africa. Only Cameroon has in its northernmost part a small share of the Lake Chad Basin and thus extends beyond the Sudan zone to the southern edge of the Sahel.

The Congo Basin is located approximately symmetrically around the equator and has a mean elevation of 400 m a.s.l. In the center, it is covered by a tropical (lowland) rainforest surrounded by wet-dry tropical Sudanese savanna. The basin is made up largely of Mesozoic to Cenozoic and recent sediments. It is surrounded by elevated Precambrian ridges. In the west, the Lower Guinea Ridge or Atlantic swell frames the Congo Basin and separates it from the Atlantic. In the north, the peneplains of the North Equatorial or Asande Ridge limit the basin. They reach a mean elevation of 650 m, extending from the Niger-Benue Basin in the west to the Wadi Bahr al-Ghazal in the east, and already lead into South Sudan. In the south, the basin is surrounded by the peneplains of the Lunda or South Equatorial Ridge. They have an elevation of 1,500–1,700 m and reach up to Angola and Zambia. The southeast of the Congo Basin includes the former province of Katanga (1971–1997 Province of Shaba) of the Democratic Republic of Congo with the former capital Lubumbashi. Since 2015, Katanga has been divided into four new provinces: Haut-Katanga with the capital Lubumbashi, Haut-Lomami with the capital Kamina, Lualaba with the capital Kolwezi, and Tanganyika with the capital Kalemie. Along the eastern part, at the border of the Western rift of the East African Rift System occurs numerous Palaeogene to recent volcanoes, with mean elevations of 2,000 m (see also Fig. 2.2).

Nowhere in Africa are there such large variances from the otherwise characteristic parallel to the latitude circle of comparable ecological features due to the climate, vegetation, and land use zones of the continent as in **East Africa**. The relief has the largest differences of elevations within the African continent itself, where elevations exceed 200 m along the margins of the rifted domal upwards of the Cenozoic East African Rift System. Climatically as well, eastern Africa occupies a special position compared to all other tropical regions on earth. The rainy seasons that recur annually or semiannually at about the time that the sun is most nearly overhead (at zenith) are overlaid by large-scale seasonal atmospheric air movements and flow, and the influence of the morphology. The first-order morphology ranges from the Afar Depression in Ethiopia, which is 116 m below sea level, to Kilimanjaro in Tanzania, the highest mountain massif in Africa with Kibo, which at 5,895 m is not only the highest mountain of the continent, but with more than 4,000 m above the surrounding savannah areas, it is also the highest freestanding mountain on the mainland earth. The relief in East Africa shows rift zones, high plateaus, and superimposed volcanic massifs, and the East African lakes result not only in strong small-scale physisogeographic features but also in a special character of the climate. The rainy **highlands and mountains** of Ethiopia, Kenya, Uganda, Rwanda, Burundi, and Tanzania are more densely populated and much better developed than the low-rainfall and sparsely populated areas below 1,000 m. Hecklau (1989) called the highlands "Inseln in einem Meer von Trockenheit" ("Islands in a sea of dryness"). The highlands are separated from each other by the Lake Turkana in northern Kenya. In the

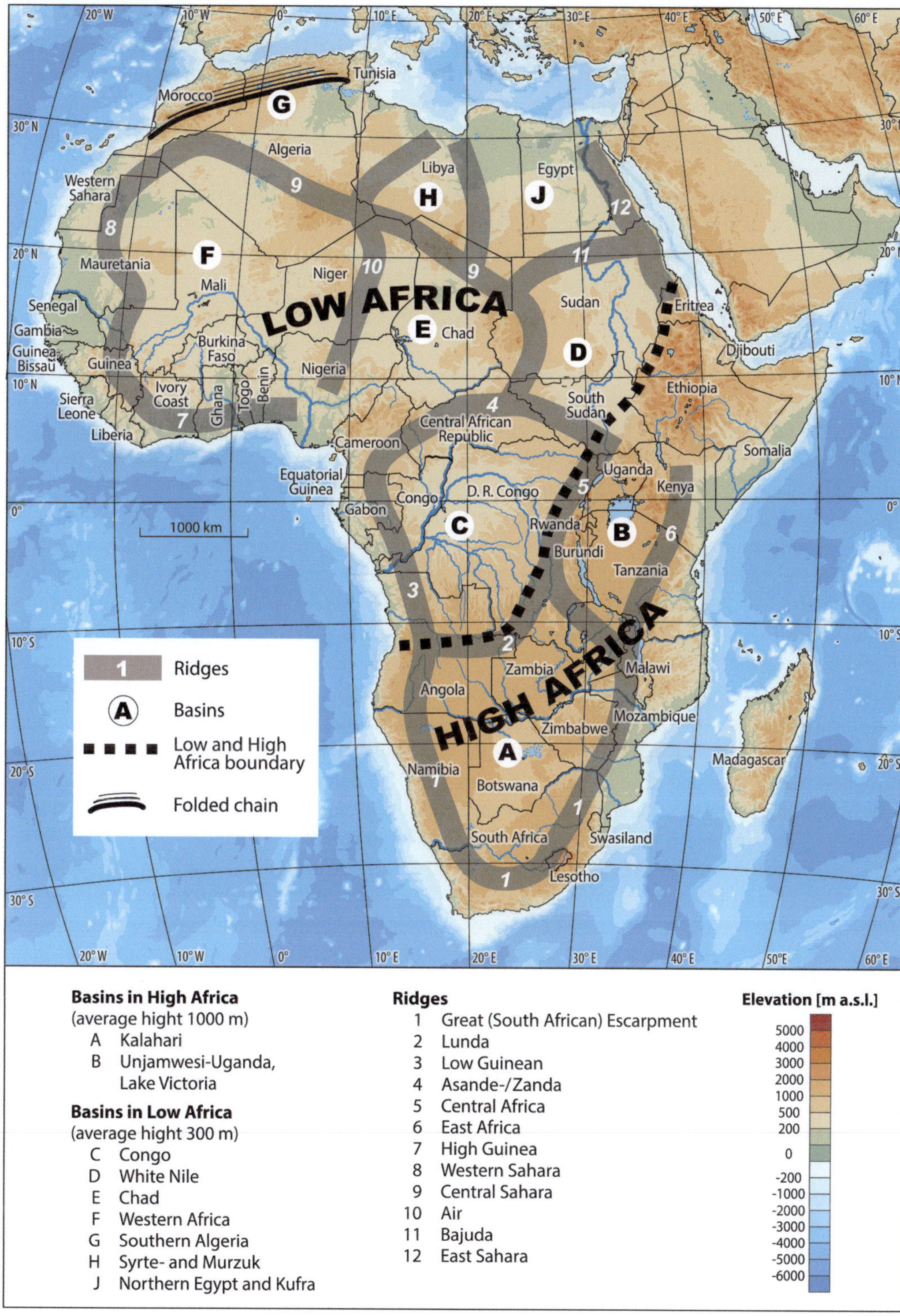


**Fig. 2.2** The geological ridges (swells) and basins of Africa

northwest, the Nile wetlands of South Sudan (Sudd) with an elevation of around 400 m bound the highlands. The narrow coastal strip of Eritrea and the Danakil Depression in Ethiopia borders the Red Sea, and with the coastal plain of Somalia, East Africa borders the Indian Ocean. In the **Horn of Africa**, Somalia, Djibouti, and the coastal area of the Gulf of Aden are some of the hottest and driest areas of the continent. In Boosaaso, located in the autonomous Somali region of Puntland on the Horn of Africa, rainfall less than 5 mm is recorded on a maximum of eight rainy days on average per year.

The **Zambezi countries** between the southern border of the Congo Basin and the South African subcontinent, as well as the Kalahari Highlands and the Greater East Africa region, do not form a geographical unit. Their geographical feature is rather a compromise, which is based on their location in the catchment area of the Zambezi River. The states of Zambia, Malawi, Zimbabwe, Angola, and Mozambique cover an area between 10° and 22° S (about 1,500 km) and between 12° and 41° E (about 3,300 km). They thus extend from the watershed between the Congo and the Zambezi, the Lunda Ridge in the north, to the northern Kalahari and the Limpopo Basin in the south, and from the South Atlantic in the west to the Mozambique Strait in the east. The physisogeographic features of the Zambezi countries, located in the wet-dry tropical climate show comparable ecological features due to the climate, vegetation, and land use zones parallel to the latitude circle. It belongs to Africa's favored landscapes. Apart from the southwest coast of Angola, where dry tropical climatic conditions occur under the influence of the cold Benguela Current as an offshoot of the Namib desert, rainfed farming is possible in all the Zambezi countries.

The vegetation is very monotonous. Thorn trees and tall grass (dry) savannas and characteristic Miombo dry forests are ubiquitous from East Angola across Zambia, Central Zimbabwe, and Malawi to North and Central Mozambique. In the southern part of the Zambezi countries (South Angola, Zambezi Valley, Central and South Mozambique), they are replaced by Mopane dry forests. The relief is characterized by the extensive, between 1,000 and 1,500 m high peneplains and plateau mountains of the Lunda Ridge and the northern Great Escarpment. In Central Angola, these form the highlands of Bihe (Bié) with heights up to 2,600 m, the Nyanga highlands in East Zimbabwe, as well as the already to the Central African Ridge belonging and the East African rift valley limiting monadnok massifs and horsts in Malawi (Mulanje mountains, 3,002 m, and Nyika mountains, 2,606 m).

South of the Zambezi countries, the Greater **South Africa** region extends with the Republic of South Africa, Namibia, Botswana, Lesotho, and the Kingdom of Eswatini (until 2018 Swaziland). The physisogeographic conditions of the greater area are determined by its altitude differentiation between the dry subtropical climate with precipitation in summer to the moist subtropical climate of the coastal lowlands of the Republic of South Africa on the Indian Ocean, the subtropical high mountain climate with precipitation in summer of Lesotho, the dry subtropical and dry subtropical climate of the Kalahari as part of a former desert with huge, currently fixed by a thorn savanna vegetation longitudinal dune fields and the Namib as a relatively narrow coastal desert, which belongs to the three most prominent deserts of Africa. The climatic division of southern Africa deviates from the typical latitudinal course of the climate zones due to the meridional distribution of moisture. As a result, the Greater South Africa region includes a wetter eastern part with Lesotho, Eswatini, the eastern part of the Republic of South Africa, and a narrow strip of the east side of Botswana, and with Namibia, West Botswana, and the central and western parts of the Republic of South Africa, a semi-arid to arid west side.

The physiogeographic factors of the pronounced climatic east–west contrast and a strong continentality are overlaid by the altitude differentiation of the relief. The approximately 1,200 m high Great Escarpment separates the narrow coastal strip on the Indian Ocean from the inland highlands, which reach heights up to 2,000 m. In Lesotho, the highlands are at 3,000 m and reach 3,482 m with Thabana Ntlenyana in the Maloti (Drakensberg). From the Great Escarpment, the central highlands slope into the Kalahari (high) Basin, whose lowest point in Botswana is around 1,000 m. To the west, the basin is separated from the South Atlantic by the Great Escarpment to the north from the Congo Basin by the Lunda Ridge with the Zambezi countries.

## References

Hecklau H (1989) Ostafrika. WBG, Darmstadt, p 572

## Further Readings

Grosz-Ngaté M, Hanson JH, O'Meara P (eds) (2014) Africa, 4th edn. Indiana University Press, Bloomington, p 376

Hagedorn H, Wagner HG (eds) (1979) Natural and economic geographical research in Africa. Würzb Geogr Arbeiten 49:328

Jürgens U, Bähr J (2002) Das Südliche Afrika. Gesellschaftliche Umbrüche zu Beginn des 21. Jahrhunderts—Zusammenwachsen einer Region im Schatten Südafrikas. Klett-Perthes, Gotha und Stuttgart

Klimm E (ed) (1985a) Harms Handbuch der Geographie. Afrika 2, die Länder südlich der Sahara. List, Munich, p 341

Klimm E (1985b) Die Sahara. In: Klimm E (ed) Afrika 2. List, Munich, pp 14–21

Klimm E (1985c) Die Guinealänder. In: Klimm E (ed) Afrika 2. List, Munich, pp 60–109

Mensching H (1969) Das Afrika-Kartenwerk 1: 1 Mio. der DFG, Bonn

Mensching H, Wirth E (1989) Nordafrika und Vorderasien. Fischer Länderkunde Bd 4:316

Petters SW (1991) Regional geology of Africa. Springer, Berlin Heidelberg

Schamp EW (1985) Mittelafrika. In: Klimm E (ed) Afrika 2. List, Munich, pp 110–151

Schiffers H (1983a) Afrika 1, allgemeine Grundzüge, Länder des Nordens. Harms Handbuch der Geographie. List, Munich, pp 13–15

Schiffers H (1983b) Blick über den Erdteil. In: Schiffers H (ed) Afrika 1. List, Munich, pp 13–23

Schiffers H, Simons P (1983) Die Natur des Erdteils. In: Schiffers H (ed) Afrika 1. List, Munich, pp 23–53

Schliephake K (1983) Der Maghreb—Übersicht über Atlas-Afrika und Nachbargebiete. In: Schiffers H (ed) Afrika 1. List, Munich, pp 157–182

United Nations (2024) Statistical yearbook, vol 67, New York, p 554

United Nations Statistics Division (2021) Geographic regions, New York. https://unstats.un.org/UNSDWebsite

Welz M (2012) Integrating Africa: decolonization's legacies, sovereignty and the African Union. University of Pittsburgh Press, Pittsburgh, p 258

Wiese B (1985a) Die Sambesi-Länder. In: Klimm E (ed) Afrika 2. (List), Munich, pp 201–232

Wiese B (1985b) Südafrika. In: Klimm E (ed.) Afrika 2. List, Munich, pp 233–258

# Geological Structure and Development

# 3

This chapter provides a general background to the key aspects of Africa's geological structure and development, forming the framework that shapes much of the continent's physical geography. Africa, like no other continent, offers unique insights into the Earth's oldest geological history. Almost 60 per cent of the continental surface comprises directly exposed **Precambrian basement complex**, dating back over 550 million (Ma) years. The geologic set-up of the remaining 40 per cent of Africa's surface is also underpinned by Precambrian basement, albeit covered by younger formations (Fig. 3.1).

Only the extreme southwest and northwest tips of Africa feature Phanerozoic fold belts. The Variscan orogeny, a geologically mountain-building phase, occurred in the younger Paleozoic 400–300 Ma years ago. This event led to the folding of the Atlas Mountains, Anti-Atlas Mountains, and Hoggar Mountains. The Mediterranean part of the Atlas Mountains in North Africa formed in the late Mesozoic 150 million years ago through the collision of the African with the Eurasian plate and the folding of the Cape Ranges of South Africa occurred 280–230 million years ago at the turn from the Paleozoic to the Mesozoic in the then southwestern border area of Gondwana. The youngest sediments in Africa include the Cenozoic volcanic and sedimentary deposits from the rifting processes in the East African Rift Valley. The Palaeogene and Neogene basin sediments, in areas like the Murzuk, Congo, and Kalahari Basins are also relatively young.

**Excursus: The Geological Timescale**

The geological time scale provides a global summary of countless small-scale temporal correlations of rock strata made at local and regional scales. It is based almost entirely upon careful observations of the distributions of fossils in time and space. Most of our knowledge of the fossil-documented Earth history record comes from the three eras of the Phanerozoic Aeon since the beginning of the Cambrian period 550 Ma ago. Thus, the Phanerozoic Aeon represents a paltry 12 per cent of Earth's history! Instead, most of Earth's history is defined by the three Precambrian eras. These older eras tell the story of Earth's beginning and life's origin. The Precambrian is further divided into the Archean (3.9–2.5 billion years) and Proterozoic (2.5 billion–550 Ma) (see Fig. 3.1). The Archean (and the Hadean >3.9 billion years) are difficult to study, however, because they are exposed in very limited places on Earth's surface and since they are the oldest eras, rocks that are Hadean and Archean in age are often buried far below younger rocks at


R. Baumhauer, *Physical Geography of Africa*, Geography of the Physical Environment,
https://doi.org/10.1007/978-3-662-71406-5_3

| Aeon | Era | Period | Epoch | Ma |
|---|---|---|---|---|
| Phanerozoic | Cenozoic | Neogene | Holocene | |
| | | | Pleistocene | 1.8 |
| | | | Pliocene | |
| | | | Miocene | 23 |
| | | Palaeogene | Oligocene | |
| | | | Eocene | |
| | | | Paleocene | 65 |
| | Mesozoic | Cretaceous | Late | |
| | | | Early | 135 |
| | | Jurassic | Late | |
| | | | Middle | |
| | | | Early | 200 |
| | | Triassic | Late | |
| | | | Middle | |
| | | | Early | 251 |
| | Palaeozoic | Permian | Lopingian | |
| | | | Guadalupian | |
| | | | Cisuralian | 295 |
| | | Carboniferous | Pennsylvanian | |
| | | | Mississippian | 355 |
| | | Devonian | Late | |
| | | | Middle | |
| | | | Early | 410 |
| | | Silurian | | 435 |
| | | Ordovician | Late | |
| | | | Middle | |
| | | | Early | 488 |
| | | Cambrian | Furongian | |
| | | | Middle | |
| | | | Early | 542 |
| Precambrian | Proterozoic | Neoproterozoic | | 1000 |
| | | Mesoproterozoic | | 1600 |
| | | Palaeoproterozoic | | 2500 |
| | Archaean | Neoarchean | | 2800 |
| | | Mesoarchean | | 3200 |
| | | Palaeoarchean | | 3600 |
| | | Eoarchean | | 4000 |
| | Hadaeum | Earth formation | | 4600 |

**Fig. 3.1** Geological time scale and exposure of Precambrian rocks, partly under younger coverage. *Source* Baumhauer et al. (2017) based on Schlüter (2008)

Earth's surface. Proterozoic rocks—which span nearly 2 billion years (42 per cent of Earth's history)—are much more accessible. The Phanerozoic is subdivided into three eras: the Paleozoic (up to 250 Ma ago), the Mesozoic (from 250 to 65 Ma ago), and the Cenozoic, which began 65 Ma ago and in which we live today. The eras are further subdivided into periods. Periods are subdivided into divisions (referred to in other languages as series)—temporally epochs—these again into stages and further into zones. A period refers to a sequence of strata, usually formed by deposition over a longer period of Earth's history, which is characterized by fossils typical of this period, known as index fossils. Magmatic rocks that have penetrated the stratigraphic sequence during the same period are also attributed to the respective period.

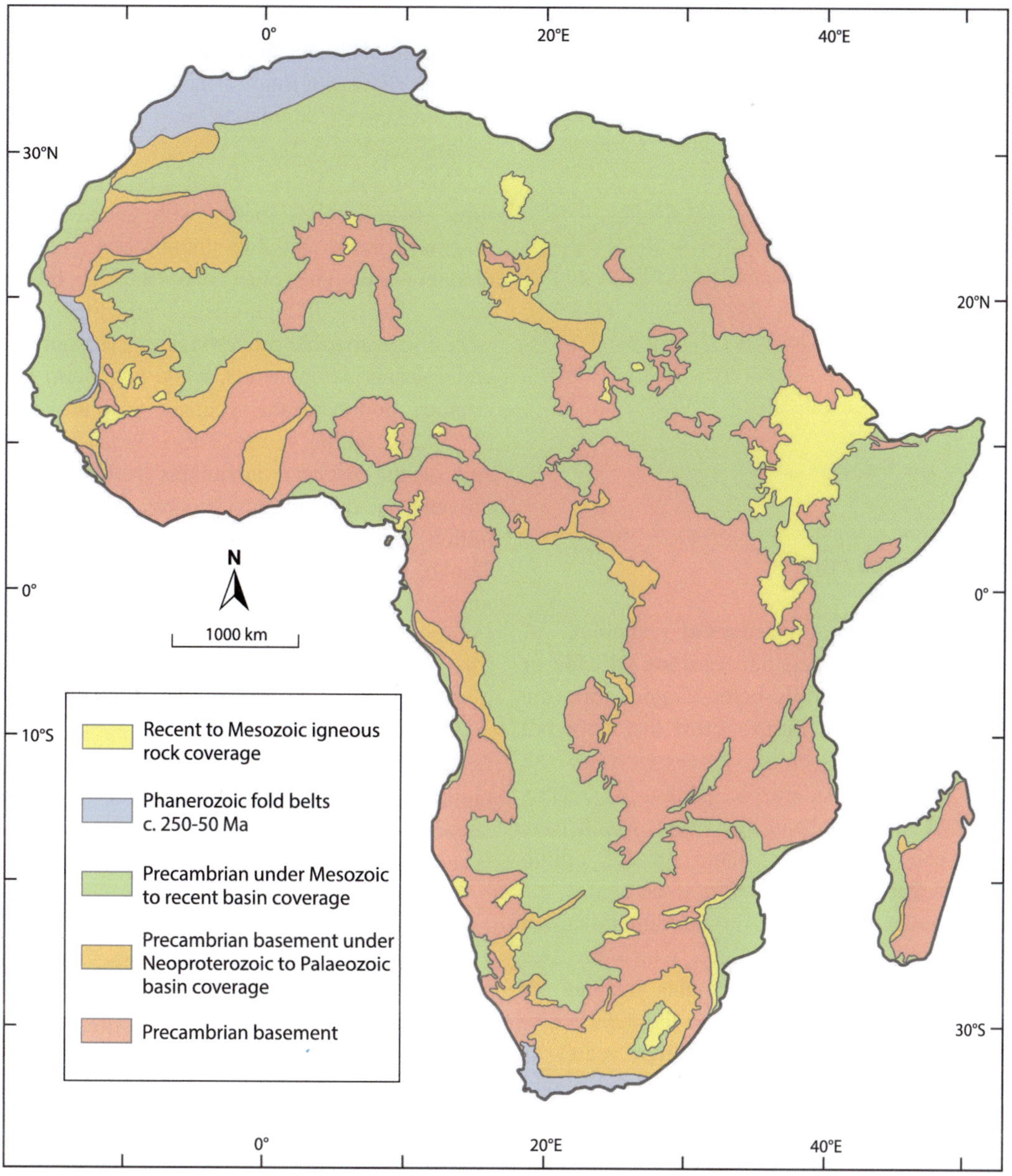


**(continued)**

The designations of the periods are often based on regions where the geological strata were first described or are particularly impressively developed (locus typicus), such as the Cambrian (Roman name for North Wales), Devonian (county in England), Permian (former Russian government in the Urals) or Jurassic (Franco-Swiss Jura Mountains).

To illustrate the temporal dimension of the development of the Earth, as represented in the geological time scale, a comparison with a calendar year is useful: The Earth was formed on January 1, and the first primitive life developed around February 21. On October 25, the development of higher life forms begins with the Paleozoic era. Since December 6, reptiles are developing while the dinosaurs already become extinct on Christmas Day. Our species appears for the first time on New Year's Eve at 11:00 pm, at 11:58 pm and 50 s, the Holocene, the youngest epoch of the Neogen (Quaternary), begins about 11,000 years ago, and the Middle Ages end a three-hundredth of a second before midnight.

## 3.1 Precambrian

The **Precambrian basement** complex of Africa can be divided into **cratons (shields** or **platforms)** and **mobile belts**. Cratons form the stable parts of the Earth's crust that have not been altered by tectonic processes since the early to middle Proterozoic. Within the cratons are even older Archean age cores, which have also been geologically stable for 2.5 billion years. A belt in geology refers to an area in the Earth's crust where mountain-building processes (orogenesis) have taken place, which involves tectonic processes such as the opening or closing of an ocean.

The **Archean** (3.9–2.5 billion years) is characterized by Greenstone Belts, and highly metamorphosed terranes. The older ones were laid down between 3.55 and 3.05 billion years and include metasedimentary components and both basal, mainly basic, volcanic rocks (komatiites), which are best preserved on the Kaapvaal Craton and its northern part, the Zimbabwe Craton. The younger ones formed 2.8–2.6 billion years ago. They are better represented in the West African Craton than in the Central African Cratons. At the base, individual volcanic cycles with basic lavas and felsic pyroclastic rocks are present. Towards the top (in geology, the hanging wall), sediments are important in the youngest greenstone belts, e.g., the Shamvaian Group on the Zimbabwe Craton, the Kambui Supergroup in West Africa, the Kavirondian Group of the Tanzania Craton, and the upper Congolese Group on the Congo Craton (see Fig. 3.2).

Greenstone Belts contain in space and time granitic series (partly with gneisses) and magmatic intrusions. Two granitic series can be distinguished. The older one (3.6–3.1 billion years) consists of high-grade migmatites (rock with distinguishable proportions of metamorphic and volcanic components), while potassium-rich granites dominate the younger granite series (2.95–2.45 billion years). They characterize the end of the Archean mountains building. This orogeny, a diachronous process, began 3.05 billion years ago in the center of the Kaapvaal Craton, took place 2.6 billion years ago in its northern part, the Zimbabwe Craton, and in the Central African cratons 2.45 billion years ago. The Limpopo Mobile Belt trends in a WSW-ENE direction for about 690 km with a maximum width of about 200 km. It separates the Kaapvaal from the Zimbabwe Craton and is dominated by high-grade gneisses. The orogenic development already began around 3.2 billion years ago and lasted until about 2.5 billion years ago.

**Excursus: Greenstone Belts and Terranes**

Greenstone Belts are zones of differently metamorphic, mafic to ultramafic volcanic sequences, which occur together with sedimentary rocks in Archean and Proterozoic cratons between granite and gneiss complexes. Their share of the cratons can be up to 20 per cent. In contrast, Terranes are made of highly metamorphosed granitic gneisses and metasediments (metamorphosed sediments). Typical metasediments include the "Banded Iron Formations" (BIF), distinctive sedimentary rock units consisting of alternating strata of iron oxides and iron-poor chert. Marble, calc-silicates, metaquartzites, coarse clastic rocks, aluminum shale, black shale, and

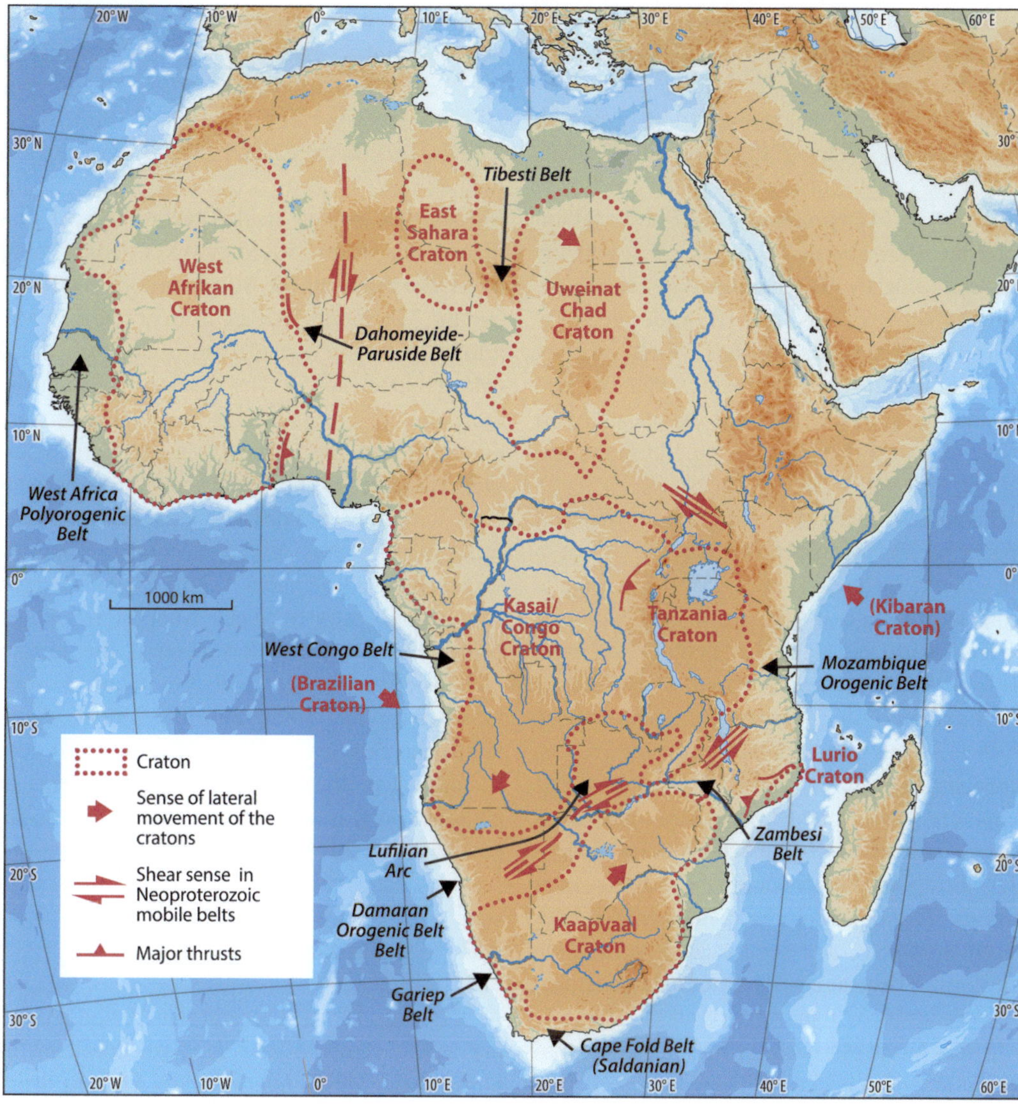


**Fig. 3.2** Assumed extension of the Archean cratons and mobile belts. From Schlüter (2008) based on Key (1992)

graywacke are also metamorphic rocks formed from sedimentary rock.

During the **Paleoproterozoic** (around 2.5 billion years ago), the Great Dyke in Zimbabwe and the Bushveld Complex in South Africa were formed by anorogenic hot-spot magmatism. In addition to the Limpopo Belt, there are other mobile belts in the marginal areas of the cratons, which were strongly metamorphosed and deformed during the late Proterozoic and early Paleozoic Pan-African orogeny. Thus, the Namaqua-Natal Belt in South Africa developed in the Paleo- to Mesoproterozoic, the Southern Mozambique Belt in the Mesoproterozoic, and the Damaran in Namibia and Zambezi Belt in Zambia and Zimbabwe as parts of the Pan-African orogeny at the end of the Precambrian. As a result of isostatic uplift and subsequent erosion, many of these Precambrian belts form the present-day morphology of Africa.

During the Paleoproterozoic period, intracontinental basins formed on the cratons. Outside the cratons, instabilities in the crust from the Paleo- to Mesoproterozoic (2.5–1.75 billion years) are shown in southern Africa due to the Eburnean orogeny. During the Neoproterozoic, orogenic activity was not as widespread as during the preceding periods. Only rifting, sedimentation, magmatism, metamorphosis, and deformation were recognized in the eastern part of central Africa (between Congo, Tanzania, and Kalahari Cratons).

Orogenic activity was limited to two main orogens in the **Mesoproterozoic**, the Kibaran Belt in the southeast of today's Democratic Republic of Congo and the arc-shaped Namaqua Province with the Namaqua Belt on the west coast of South Africa. A large proportion of the Kibaran Belt comprises metasediments, which likely exceed 10,000 m in total thickness. Intrusions exist in the form of granitic gneiss complexes and highly complex granitoids (composite granitoids). The Namaqua Province is lithologically much more complex with basement gneisses, supracrustal rocks, syntectonic shield-shaped intrusions, and small-scale post-tectonic intrusions. The province is extensively mantled by Neoproterozoic to recent deposits.

In the **Neoproterozoic**, widespread tectono-thermal activities occurred. They are recorded at about 950, 785, 720, 685–660, and 600–450 Ma. The lithological components are characterized by clastic and chemical sedimentary rocks with fluvio-glacial deposits and limestones, for example, in the Volta and Togo Belt in West Africa or by the limestone and quartz group in Morocco. Volcanic rocks form either in alternation with powerful sedimentary sequences or as intrusive rock sequences. Intrusive rocks of the alkali and calc-alkali granitoid series, including syn- and post-orogenic intrusions of granodioritic rocks, e.g., in the Mozambique Belt in Kenya. Major pegmatites are common, e.g., the Khan-Pegmatite of Namibia.

In the **Late Neoproterozoic**, the most extreme global glaciations of Earth's history occurred, which also affected Africa. Numerous glacial deposits from the late Proterozoic (tillites, diamictites, etc.) and morphological indicators (polished bedrock surface, Roche Moutonnée, tension cracks) record at least four glaciations from 750 to 580 Ma. For two of these, the **Sturtian Ice Age** from 715 to 680 Ma and the **Marinoan Ice Age** from 660 to 635 Ma, a total glaciation of the Earth is even suspected (Macgabhann 2005). The glaciers spread from the poles to near the equator, the oceans were largely frozen, and nearly the entire Earth's surface was covered with ice. Three causes of these early Ice Ages seem possible. First is a reduction in the sun's energy output. Second is a rapidly decreasing $CO_2$ content. Third is an increase in tectonic activity. This Neoproterozoic or Neocambrian ice age has entered the history of science under the term "**Snowball Earth**" (see Kirschvink 1992; Hoffmann et al. 1998; Hoffmann and Schrag 2002). However, this geo-scientific hypothesis, based on a few pieces of evidence only, is still highly controversial today (Brooks et al. 2016).

## 3.2 Paleozoic

At the close of the Precambrian period, Africa lay at the center of the southern continent of **Gondwana.** Throughout the **Pan-African orogeny** (also referred to as the Pan-African event) 950–450 Ma ago, Gondwana became part of the emerging Supercontinent **Pangaea** and was structurally (again) divided into cratons and mobile belts. The Pan-African orogeny began with an initial rift phase, sedimentation, and magmatism. The opening of the ocean with the genesis of the Iapetus Ocean, and continental margin sedimentation as well as subduction and plate collision with subsequent magmatism are the characteristic features of these plate tectonic processes and the structures associated with them, which are found in the Pan-African orogeny. Throughout most of the Paleozoic times, North Africa occupied the southern seaboard of the Iapetus Ocean, whereas southern Africa was bordered by a shelf sea to the south.

In contrast with the Archean cratonic areas, where only kimberlite bodies are common, the successive Proterozoic mobile belts have been the sites of Phanerozoic faulting, tectonics, and magmatism since the end of the Pan-African orogeny. Anorogenic magmatism is widespread in Africa, always associated with rift formations and faults/fractures and since the Mesozoic also with the formation of the **East African Rift System.**

With the beginning of the **Cambrian**, the sea extends over large parts of Northwest Africa, such as Morocco, the West African Sahara, Mauritania, Senegal, and Guinea. In Morocco and Guinea, tillite (consolidated moraines) is included in the Early Cambrian. In Central Africa, there are no signs of the presence of Cambrian strata except in the northern Kivu region of the DR Congo. In the central and eastern Sahara, marine sandstones extend from the Hassi Messaoud Basin (as part of the South Algerian Basin) in southeast Algeria over the Tibesti Mountains to western Egypt. The approximately 500 m thick Hassi-Messaoud sandstone has continental facies (Fig. 3.3).

The **Ordovician** in Africa is formed as a sandstone sequence that can locally reach several hundred, in the Anti-Atlas over 1,200 m. In Algeria, the comparable strata are metamorphosed. However, in Morocco, there is also a complete, fossil-proven Ordovician with graptolites and trilobites. The Ordovician of the Sahara is characterized not only by three erosion gaps but also by a new tillite stratum. Further occurrences of Ordovician are from the Adrar in Algeria, the Fezzan, and from the eastern framing of the Murzuk Basin in Libya, as well as from the Tibesti Mountains in Chad and the

**Fig. 3.3** Late Cambrian and Ordovician sandstones on the Djado Plateau at the SW-margin of the Murzuk Basin, central Sahara, Rep. of Niger. *Photo* R. Baumhauer

Jebel Uwainat located right on the border of Egypt, Libya, and Sudan. The greatest thicknesses of Ordovician sediments were, however, drilled in the Syrte and Murzuk Basins of the central Sahara, where the Ordovician Sea has deposited sediments over 3,000 m thick. The Guinea plateau has a 500 m thick sandstone depositional sequence, which does not provide body fossils itself, but due to the underlying and overlying, fossil-bearing Cambrian and Silurian sequences can also be assigned to the Ordovician. In other parts of the continent, Ordovician is absent, except for the Late Ordovician of the Table Mountain Group Sandstone in South Africa.

The **Silurian** in North Africa is a regional-metamorphic transformation. In Tunisia and Tripolitania in northeastern Libya, graptolite shales up to 500 m thick have been found. Silurian of comparable thickness and formation also occurs in the Sahara, in the Hoggar Mountains of Algeria, in Fezzan, in the Murzuk Basin, and in the Tibesti Mountains. Fossiliferous Silurian is widespread in Guinea (Early Silurian as graptolite shale, Late Silurian as cephalopod limestone). In the Late Silurian of the Central Sahara, terrestrial and lagoonal to deltaic sandstones are deposited during a marine regression period.

The **Devonian** in North Africa is lithological as well as faunally very distinct from southern Africa. In the Anti-Atlas, the Middle to Late Devonian is present in a 3,500 m thick depositional sequence of shale and limestone. To the east, marine Devonian covers the Sahara and the Hoggar Mountains as well as Fezzan; the Murzuk Basin, the Tibesti Mountains, and Ennedi in Chad. The rocks are characterized by continental sandstones with Devonian plant remains. Marine Devonian is also found in Guinea and on the coast of Ghana. In the Cape Province of South Africa, the 800 m thick Bokkeveld Group has yielded various marine fossils of the Devonian age. In the latest Devonian, regression conditions are established throughout Africa.

Marine **Carboniferous** is limited to North Africa (Morocco and Algeria on the one hand, Egypt on the other). This includes the marine horizons in the coal basins of Jerada (Morocco) and Bechar (Algerian Sahara). Marine Carboniferous then reappears in eastern Egypt on the Sinai Peninsula on the Red Sea. In the Late Carboniferous, the sediments are represented by a continental sandstone sequence, marking the lower part of the "Nubian Sandstone". In general, marine deposits of the Permian and Triassic in Africa are extremely rare. The Permian has yielded some localities of marine deposits in southern Africa but less in North Africa, as an offset of the Tethyan Ocean in southern Tunisia and Libyan Tripolitania.

## 3.3 Mesozoic and Cenozoic

The Late Carboniferous to Early Jurassic period in sub-Saharan Africa is represented by widespread non-marine strata, belonging to the **Karoo Supergroup.** The geological term "Karoo" originally being used in a generic sense for tectonically and climatically controlled continental sediments and sediment structures ranging from the Late Carboniferous to Early Jurassic of the Karoo Basin in South Africa. With advancing knowledge, the term was applied to all continental sediments that emerged from the extensional basin structures of Gondwana. Comparable sediment basins were also found in India, Antarctica, and Australia. The Karoo series covers about 600,000 km$^2$ in Africa and reaches thicknesses up to 7,000 m (see also Figs. 3.4 and 3.7).

A typical Karoo stratigraphic succession (see Fig. 3.5) comprises from the base to top tillites, coal seams, and fan-shaped clastic wedges, which are interlocked with lake deposits, river or aeolian sediments, and widespread basalt flows. The basal Dwyka Group belongs to the Late Carboniferous and comprises a basal sequence of shales and sandstones up to 400 m in thickness, succeeded by the more than 400 m thick glaciogenic Dwyka Tillite, characterized by block packings, scratched boulders over glacially scoured rock substrate, and coastal inclusions of marine strata as well as finally a hanging sequence of alternating glacial and

**Fig. 3.4** The Nuweveld piedmont scarp 40 km north of the city George in the South African province of Western Cape made by Permian and Triassic fluvial sediments of the Beaufort Group (Karoo Supergroup). *Photo* University of Witwatersrand/CC

interglacial deposits, which have yielded floras with Lepidodendron and Gangamopteris. Lepidodendron is a genus of the "scale trees", tree-shaped club mosses, which were an essential component of the Carboniferous coal forests; Gangamopteris is a fern characteristic of the Late Carboniferous-Early Permian glaciation of the southern continents. The Dwyka Tillite with its outcrop width of 2,250 km around the Karoo plateau, most clearly marks the sustained Late Carboniferous-Early Permian glaciation of South Africa. This extensive inland glaciation features a general south to southwest direction of ice movement, radiating from at least four major ice centers (Namaland Ice, Griqualand Ice, Transvaal, and Natal Ice). This inland glaciation is part of the **Late Carboniferous-Early Permian glaciation** of the southern continents and the Indian subcontinent.

Eye-catching is the close correspondence of a 50 m thick bituminous shale series with Mesosaurus remains ("White band") in the upper side of the Dwyka Tillites of South Africa, which recurs in completely analogous formation in the Irati strata of the Parana Basin of South America (also as bituminous shale with Mesosaurus). A coal-bearing clay shale-sandstone sequence between Dwyka and the "White band" recurs in a similar formation (as coal-bearing siltstones) in the Parana Basin (Fig. 3.6).

The Dwyka Group is succeeded by the coal-bearing alluvial to flysch breccia of the Ecca Group, which in turn is overlain by deltaic wedges of the Beaufort Group that pass upward into fluviatile and eolian deposits (Molteno, Elliot, and Clarens Formations). Voluminous coatings of the lava flow of Drakensberg basalt in the Jurassic, referred to as the Stormberg Group, with plant-bearing clay shales and sandstones, red sediments with reptile tracks, cavernous light sandstones with silicified tree stumps as well as remains of fossil dinosaurs and crocodiles, together up to over 1,000 m thick, ended the Karoo depositional cycle.

The main Karoo Basin extends in an E-W-direction across southern Africa and subsided as a foreland basin because of prolonged regional compression and upwarp in the Cape Fold Belt. Outside of the Karoo foreland basin are shallow, wide intracratonic subsidence troughs (sag basins) to the west. These sag basins are known in the Republic of South Africa, Botswana, Namibia, Angola, the DR Congo, and Gabon.

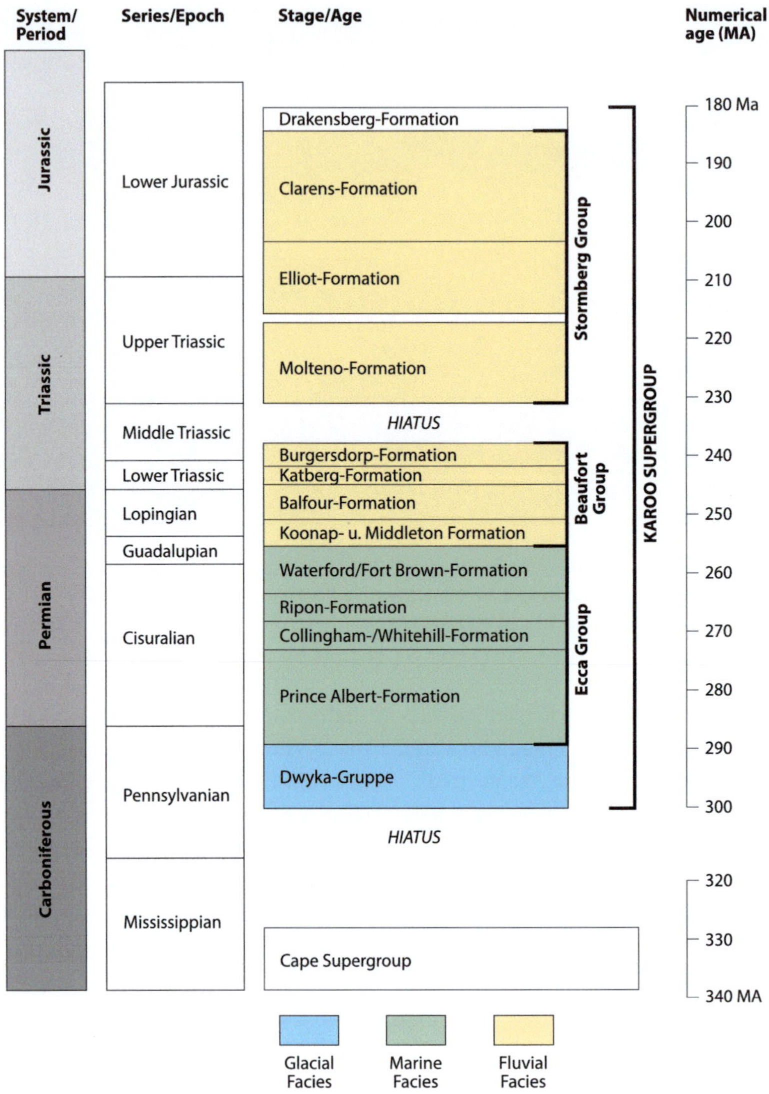


**Fig. 3.5** Karoo Supergroup. Modified after Catuneau et al. (2005)

The easternmost group of the Karoo Basin is represented by narrow rifts and half-grabens or troughs. They lie in eastern and southeastern Africa in Tanzania, Kenya, Uganda, Zambia, Zimbabwe, and Madagascar. The beginning of the sedimentation cycle that led to the formation of the Karoo sequence already took place during the times of Pangaea and shows that the Phanerozoic surface formation of southern Africa was no longer determined by extensive orogenesis, but by an alternating cycle of deposition and erosion phases. Today, the basement is widely exposed, especially in upwarp zones (swells, ridges) that frame the deeper sedimentary basins. Typical examples can be found in the basin areas of the Zambezi, Congo, Niger, and Nile. In the older basins, marine sediments (calcites, dolomites, and shales) are present, while in the younger basins, Late Mesozoic shales and sandstones were deposited (Fig. 3.7).

In North Africa, marine fossils of the **Triassic** are recognized at the beginning of the Middle Triassic around 250 Ma (in Algeria, Morocco, Tunisia, and the north of Libya with

**Fig. 3.6** Glacial striations of the Late Carboniferous–Early Permian glaciation in the north of the Rep. of South Africa. *Photo* R. Baumhauer

dolomitic limestones and gypsum-bearing, colorful marls with thicknesses up to 1,000 m over continental Lower Triassic) and in the Late Triassic over lagoonal or continental lower Late Triassic. In Egypt, the 200 m thick Middle Triassic sequence of the northern Sinai Peninsula continues in depth to the area of the Gulf of Suez and Cairo. Triassic marine fossils are unrecognized in the rest of northern Africa. This period marks the beginning of a continental epoch lasting until the end of the Early Cretaceous. The typical succession of sediments, "**Continental intercalaire**", comprises sand or sandstone sequences without absent differentiation of stratigraphy.

**Pangea's breakup** and thus the "creation" of today's Africa began about 250 million years ago. Marine Permian on the west coast of Madagascar shows that the Mozambique Channel also opened at this time. In the Early Triassic, Pangaea split into two mega continents, the northern continent **Laurasia** and the southern continent **Gondwana**. In the Early Jurassic, the Indian subcontinent, Australia, Antarctica, and Madagascar separated from Gondwana, which broke into the two continents South America and Africa. However, it took over 100 Ma until Africa reached its today form and position. Since the breakup of the Arabian Peninsula and the development of the Red Sea in the Miocene epoch of the Paleogene period about 15 Ma ago, Africa is surrounded by oceans on all sides. Since the African plate, except in the north, is bounded by mid-ocean ridges, it is constantly gaining area. At present, the African plate is continuously moving northeast at about 2 cm per year.

The breakup of Gondwana in the Mesozoic and thus, the collision of the proto-African plate and the South American plate caused the folding of the South African **Cape Fold Belt**. The orogeny is completed by the beginning of the Cretaceous. Simultaneously with the breakup, intensive volcanic activity occurred.

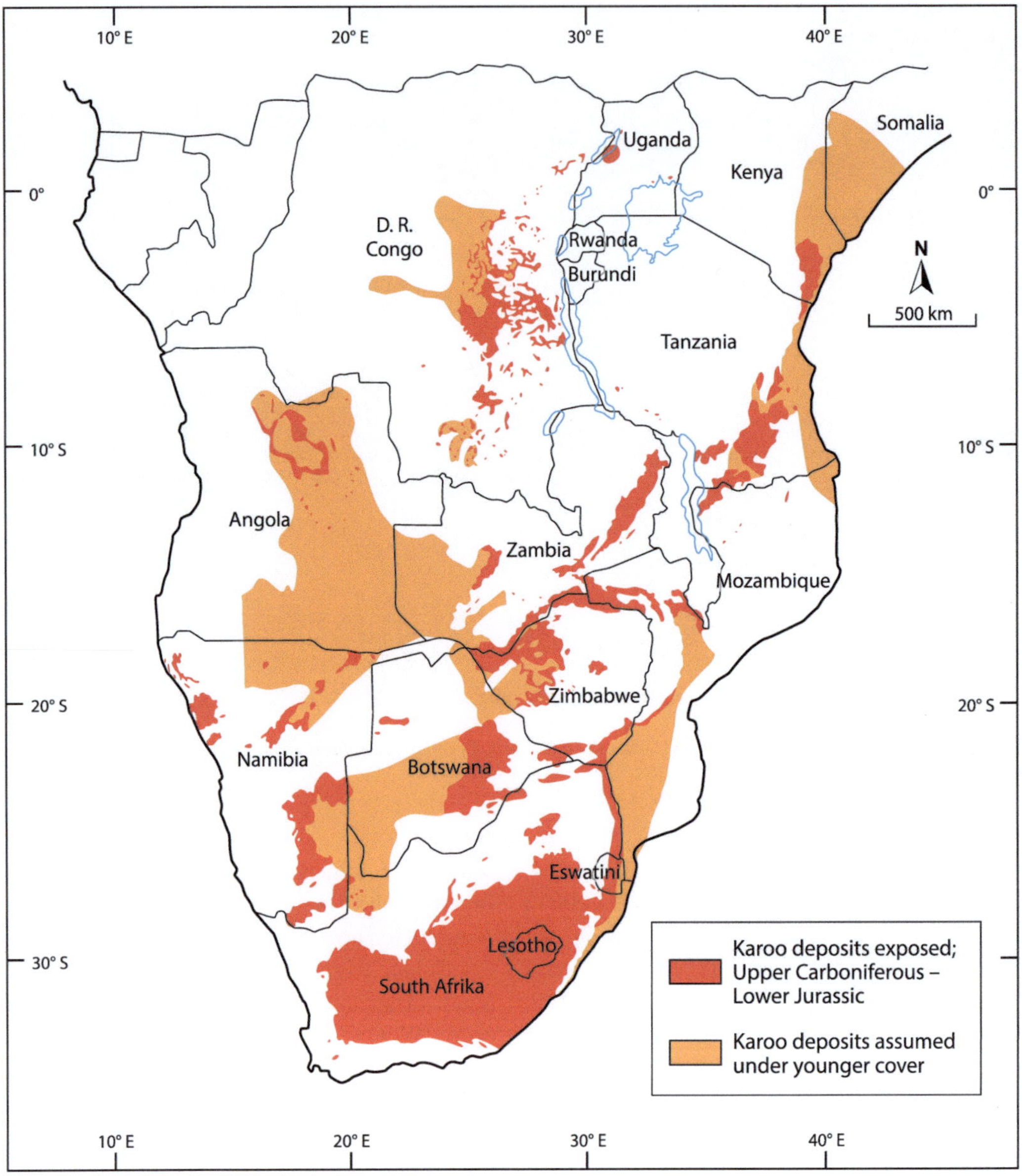


**Fig. 3.7** Deposits of the Karoo Supergroup in southern and eastern Africa After Schlüter (2008)

In parts of NW Namibia, Botswana, and the Drakensberg of South Africa, the remnants of this old relief rejuvenation are exposed on the surface. They cover approximately 140,000 km$^2$. The present landscape of Africa is very much a product of the breakup of Gondwana, and the tectonic processes associated with this event. The other major influence since the creation of Africa as a distinct continental block has been the continuing episodes of rifting and volcanism that have occurred. They have to account for certain specific characteristics of present-day Africa. One notable feature is a series of broad upwarps that rise to altitudes of several hundred meters above the surrounding terrain, covering much of the continent. These are particularly evident as the long-recognized marginal upwarpings,

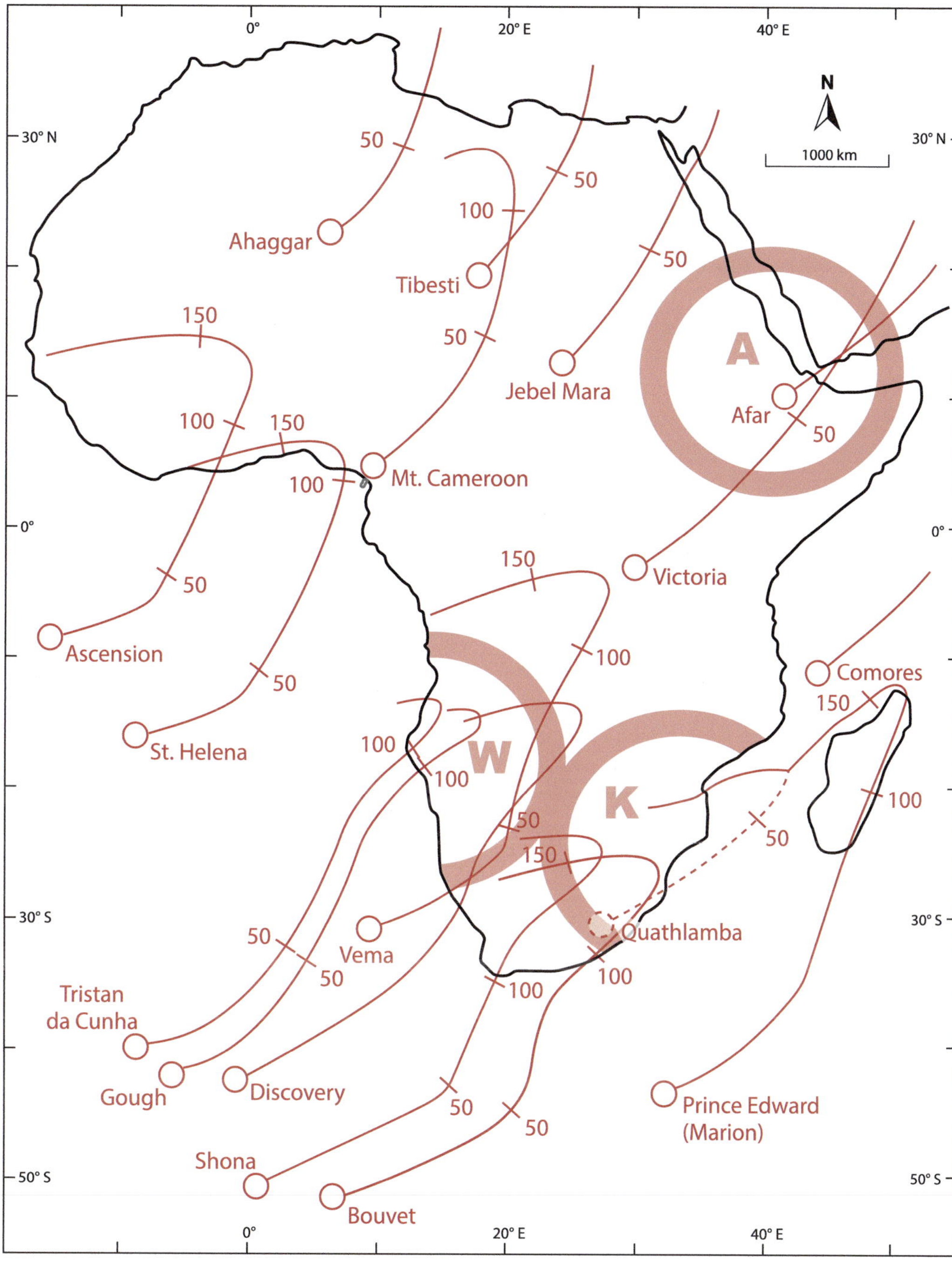


**Fig. 3.8** Africa in relation to hot-spot tracks (solid lines labelled at 50 Ma age increments, present hot-spot locations are indicated by small circles) and assumed mantle plumes: A = Afar, Karoo, W = Walvis. Modified from Summerfield (1996)

running parallel to the coastline in several areas and flanked seaward by a sharp topographic discontinuity in the form of escarpment/s. The most prominent upwarpings, however, are those associated with the much younger East African Rift System (Fig. 3.8).

Two volcanic regions are associated with the breakup (geological: rifting) of Pangaea in the Mesozoic: the Karoo flood basalts in Southeast Africa and the Etendeka lavas in Namibia. The Cenozoic flood basalts of Ethiopia are linked to the significantly younger Arabian Peninsula breakup, the development of the Red Sea, and thus a new continental rift, accompanied by complex graben/half-graben structures. The **East Africa Rift System** is thus the result of ongoing continental fragmentation processes through extension tectonics. They extend in two main structures from north to south in the eastern half of the continent, from Jordan and the Red Sea to South Africa. It is assumed that a mantle plume or a hotspot is located under the rift, causing a bulging and weakening of the lithosphere. The results from this include the upwarping of the former land surface, faults along or near the crest of this elevated landscape, and in some regions volcanism. Overall, the East African Rift System can be divided into regions with high magmatic activity (Afar Depression, Ethiopian Rift, Gregory Rift) and with low activity (Albert, Tanganyika and Malawi Rift Zones). Dominant magmatic rocks are alkali basalts, but phonolites, trachytes, and rhyolites also occur (Fig. 3.9).

Another rift system in Africa outside the East African Rift Zone forms the **Cameroon Line**. It was formed in the Cretaceous period along a rift with pronounced volcanic activity and extends from the Mandara Mountains in northern Cameroon, the Atlantika Mountains, the Adamawa Highlands, the Bambouto and Manengouba Mountains via active volcano Mount Cameroon to the islands in the Gulf of Guinea. Along the Cameroon Line are several volcanoes that have been active since the Cenozoic and Neogene.

The most volcanically active area of Africa is to be found at the Great East African Rift Valley, but there are also volcanic deposits north of the Cameroon Line in the central Saharan mountains Hoggar, Tibesti and Aïr (Fig. 3.10). These deposits

**Fig. 3.9** The volcanic Kibo, at 5,895 m, the highest peak in Africa, is part of the Kilimanjaro massif, which is located on the eastern shoulder of the East African Rift System. *Photo* S. Auth, Mauritius

**Fig. 3.10** The volcanic Ilamane (2,739 m) in the Atakor massif of the central Hoggar Mountains, Central Sahara, South Algeria. *Photo* R. Baumhauer

cover the Precambrian basement complex. The development of Hoggar and Tibesti shows a post-Cretaceous uplift. The relief of the southernmost part of the continent is also a product of uplift, volcanism, bulging, and rift formation. The Karoo deposits were widespread formed by extensive flood basalts, sandstones, and shales, and different upwarping rates, bulging, and erosion affected the extensive area of uplift of the old Cape Fold Belt and the Drakensberg.

In the Late Neogene (**Pleistocene**), the Precambrian basement was extensively exposed in West Africa, overlain only by thin, fine-grained deposits of clayey-silt and sorted sediments from flowing waters. The thickest accumulations of these deposits occur in the Lake Chad Basin. The most recent geological development of Africa is characterized by sustainable changes in the water system: Even in the Pliocene and Early Pleistocene, numerous rivers from the marginal mountains of the continent flowed into the originally endorheic intra-African basins, in which extensive freshwater lakes formed. In present Africa, almost all these basins drain exorheically (see Chap. 6).

In East Africa, the landscape is dominated by rift valleys from Ethiopia to Mozambique with large lakes and powerful Cenozoic fluviolacustrine deposits as well as volcanic debris. The rift structures form a linear feature, but individual basins are oriented at highly variable angles to the overall trend of the rift system. The relationship of Late Cenozoic rifts to earlier structures varies; in some locations such as Lake Malawi, the Late Cenozoic basin cuts across two Permian–Triassic basins, but in other cases Late Cenozoic extension has involved the reactivation of much earlier structures, e.g. the southern half of Lake Tanganyika overlies part of the Permian–Triassic Luama Rift.

During the continental Late Cenozoic (Quaternary) of Africa, severe oscillations in climate and vegetation were experienced as on other continents. Vegetational zones contracted and expanded in response to changes in precipitation and temperature. The extent, phase, and duration of wet to arid changes in Africa are correlated in many cases with glacial to interglacial environments of Eurasia. However, they varied extensively from region to region. Finally, the Pliocene–Pleistocene deposits also provide significant hominid finds (see Chap. 5).

## 3.4 Natural Resources in Africa

Nearly half the world's gold, up to 90 per cent of its chromite and platinum metals, and one-third of all minerals are in Africa. According to the United Nations, Africa is home to about 30 per cent of the world's mineral reserves, 12 per cent of the world's oil and 8 per cent of the world's natural gas reserves. The deposits of **Bauxite, Chromite**, **Cobalt**, **Coltan**, **Diamonds**, **Manganese**, **Phosphorite**, **Titanium minerals**, **Vermiculite**, and **Zircon** are located at the leading exploration sites on a global level (Fig. 3.11).

Three geological formations are important for the most significant natural resources: The Precambrian basement complex, for example, provides Copper and Zinc on the Lunda Ridge, Gold in the eastern Democratic Republic of Congo on the Central African Ridge, or Uranium in Niger on the Aïr Ridge. The Paleozoic Karoo Supergroup succession has partly rich coal deposits, and in the Cretaceous to Cenozoic sediments in shelf areas, large hydrocarbon deposits have been found. The North African oil deposits in Algeria and Libya, on the other hand, are linked to Paleozoic basin sediments. In particular, southern South Africa and southern Central Africa are focal areas of the economically significant mining industry, but also, in West Africa, the mining sector occupies an important economic position.

However, the large mining areas and the most important isolated deposits for metals are all to be found in the ridge areas of High Africa. The sediment basins on the plateaus, for example, the Kalahari Basin, are largely free of deposits. Leading mining areas in High Africa are Witwatersrand, the Bushveld Complex and the Limpopo Province in the Republic of South Africa, Great Dyke in Zimbabwe, and the Copperbelt in Zambia and in the DR Congo. In the Kibaran Belt in Rwanda, Burundi, Western Tanzania, and Southwest Uganda are economically important occurrences of tin, beryllium, tungsten, niobium-tantalum, gold, and lithium.

A quarter of the world's mined **gold** and more than half of the globally mined **diamonds** come from sub-Saharan mines. And more than half of the world's gold deposits are located in the Archean rocks of South Africa. In the Transvaal region of the Republic of South Africa, the sediments of the Precambrian Witwatersrand Basin near Johannesburg have delivered more than 30,000 tons of gold and also lead in terms of the world's uranium production. Mineralizing fluids concentrated the auriferous ores by migrating through conformable passageways created by tectonic movements. They are bound to the coarsest grain sizes and are found in channel-like fillings of the basin edge, and turn into non-exploitable rocks towards the basin. In addition to gold, byproducts such as silver, copper, arsenic, bismuth, and pyrite are mined.

Gold is found almost in West Africa and was successfully mined already in historical times. In Nigeria, throughout the western and north-western parts of the Nigerian basement, gold deposits are mined. Within the granites of the Jos Plateau and the region around Abakaliki, large tin and lead deposits have been depleted since pre-colonial times. Gold deposits are economically by far the most important mineral type in Ghana. Alluvial gold concentrations in recent and subrecent river gravels (secondary deposits as placer or wash gold) of the archaic rocks of the Upper Guinea Ridge, attracted Europeans in the eighteenth and nineteenth centuries to the named after it Gold Coast of today's Ghana. The first gold mine built and exploited by Europeans was established in 1878 in Tarkwa, Ghana. Prior to this, Europeans

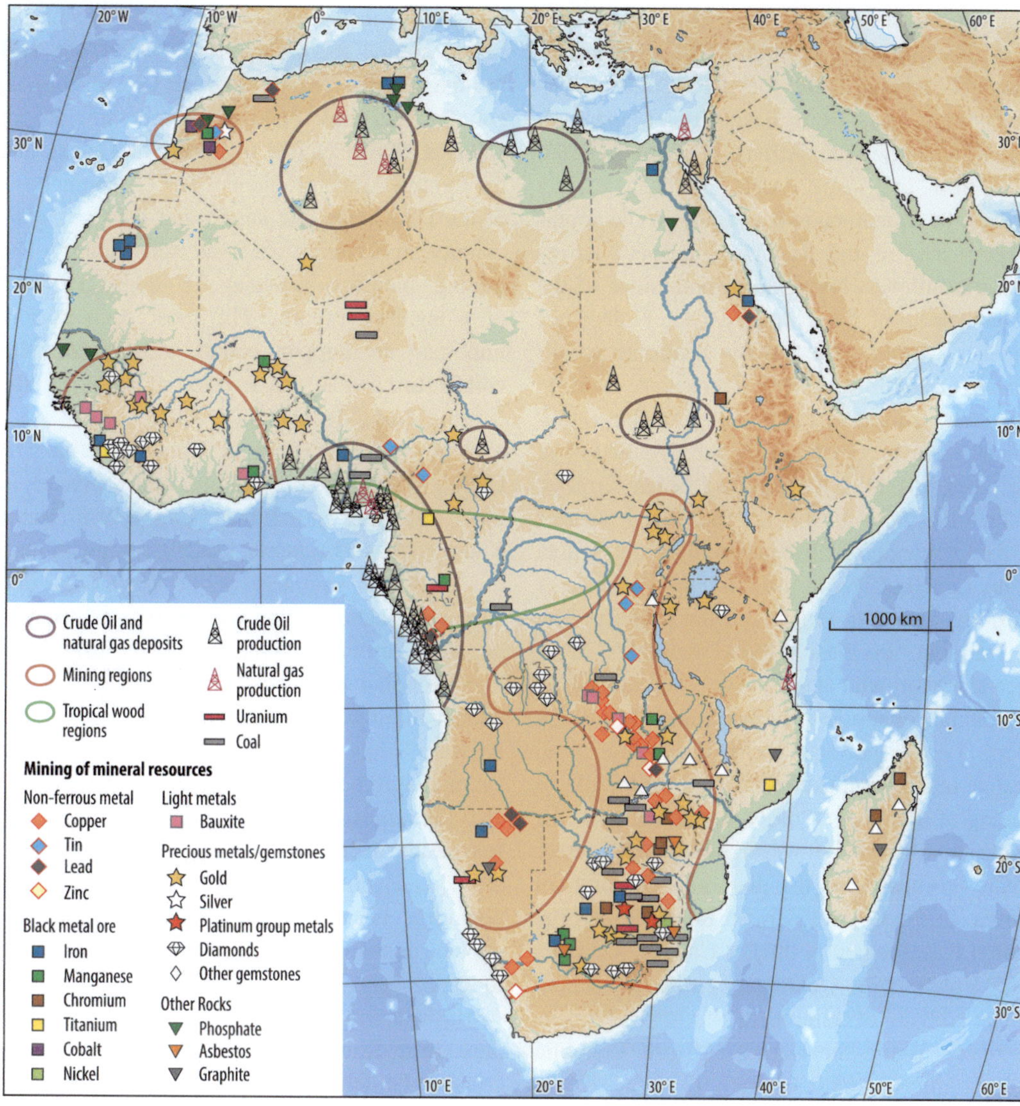


**Fig. 3.11** Natural resources in Africa

acquired gold through trading stations, purchasing it from local traders. To this day, gold deposits remain the most economically significant mineral resource in Ghana. With a production of 135 tons (2023) and reserves of around 1,000 tons, it is the most important industrial gold producer in West Africa and the largest producer in Africa.

The only primary source of **Diamonds** in southern Africa is kimberlite, an ultrabasic igneous rock that occurs as volcanic pipes (kimberlite pipes), dykes, and sills. Most of them are Jurassic to Cretaceous in age. One of the most famous diamond deposits is located near Kimberley in South Africa. The diamonds are found in the fine-grained matrix of volcanic breccias, often with carbonate binders. The diamond content decreases with depth, and mining is worthwhile up to about 1,000 m depth. The pipe diameter averages 800 m. The largest mining areas are in southern Africa and southern Central Africa (Rep. of South Africa, Botswana, Namibia, Angola, and the DR Congo), further mining areas are in East Africa (Tanzania), in western Central Africa (Central African Republic), and West Africa (Ghana, Sierra Leone). In Namaqualand, a region in the southwestern part of the Republic of South Africa and in Southwest Namibia on both sides of the lower course of the Orange River, kimberlites together with melilite basalts penetrate fossil-bearing strata of the Lower Cretaceous. Starting from the primary occurrences, the placer diamonds were accumulated by the denudation of the pipes in fluvial terraces and beach terraces. Known Examples of this are the fluvial terraces of the Oranje and its tributaries, the estuary area of the Lepelle River in the Limpopo Province of the Rep. of South Africa, and the Lüderitz Bay in Namibia. In West Africa, diamonds, from secondary, alluvial deposits only, are found in exploitable quantities in Ghana and Guinea. In Sierra Leone, political turmoil caused a total decline in diamond exports, except for the illegal trading of diamonds ("blood diamonds") for the provision of arms. However, they are of lesser quality compared to the diamonds of southern Africa. In the Central African Republic and the central Kasai region of the DR Congo, diamonds are depleted in Quaternary alluvial deposits, as well as by mining of the lower terraces of rivers. While the Central African Republic still produced 468,000 carats (1 carat, Kt or Ct = 0.2 g = 200 mg) in 2007, production fell to 11,157 carats in 2020; the DR Congo produced 8 million carats in 2023, making it the second-largest producer of natural diamonds (jewelry and industrial diamonds) after Russia. Since it is possible to extract diamonds without modern technological aids, their mining is largely carried out in so-called artisanal small-scale mining, where the diamonds are mined by hand using the simplest technical means, sometimes up to 15 m below ground level. It is assumed that more than one-third of the country's diamond production is smuggled out each year.

The most important mineral deposits of Central Africa can be found in the granite and gneiss rocks and metasediments of the Precambrian ridges around the Congo Basin: **Iron ore and manganese ore** in Gabon and DR Congo, **copper, cobalt, zinc, and tin** in the former province of Katanga (DR Congo), **uranium** in Gabon and the Central African Republic. The economic basis of the DR Congo, the largest and most populous country in Central Africa, is still mining. The mining industry accounts for 80–95 per cent of Congolese exports. The most economically important are **copper, cobalt, diamonds, gold, coltan, tin**, and as was revealed in 2019, also the largest **lithium deposit** in the world.

The former province of Katanga in the DR Congo, which was divided into four provinces in 2015 following an administrative reform (Haut-Katanga, Haut-Lomami, Lualaba, Tanganyika), lies in the far southeast of the DR Congo and is a world-class mining region. It is referred to as the **Copperbelt** which extends southward into Zambian territory. It is the second-largest copper mining area in the world after the South African Transvaal/Rand region, and with an estimated 70 per cent of global production, it is also the largest cobalt mining area in the world. It extends over 300 km with a mean width of 50–80 km. The sediments of the Copperbelt contain near-surface, open-pit mineable copper deposits with a copper content of up to 6 per cent. Around 20 per cent of today's world reserves of copper ores can be found here. In addition to copper, byproducts such as cobalt, platinum, lead, zinc, and manganese are depleted.

The Copperbelt of (former) Katanga (today Haut-Katanga, Haut-Lomami, Lualaba, Tanganyika) within the geological Lufilian Arc at the eastern end of the Lunda Ridge is a polymetallic metallogenic province, the major types being stratiform copper deposits formed by weathering, in which the dominant mineable deposits are Cu–Co and Zn–Pb sulphides, Cl oxides and noble metals. Mines in the Copperbelt are Lonshi in the south, Tenke Fungurume in the north, and Shinkolobwe slightly south of it. Other copper mining areas are in Botswana, South Africa, and Zimbabwe.

Although many multinational mining companies have settled in the DR Congo, only a few jobs have been created because of a lack of skilled workers. In contrast, both in the DR Congo and in Zambia, non-industrial mining takes place on a large scale. Up to one million people are said to work in so-called artisanal small-scale mining. Without adequate labor and social and environmental standards, this small-scale mining is associated with risks such as child labor or the financing of violent conflicts and human rights violations. On the other hand, the artisanal mining activities are important for the international raw materials supplies. A good example is columbite-tantalite, better known as coltan. It is found in major quantities in the eastern part of the DR Congo. When refined, coltan becomes (heavy) metallic tantalum, a heat-resistant powder that can hold a high electrical charge. It is used as a technology metal in smartphones and other modern electronic devices. The DR Congo supplies 30 per cent of the worldwide tantalum production. There, the high-melting metal is almost exclusively mined in artisanal mining. The Congolese artisanal miners also contribute significantly to cobalt production: 10 per cent of the world's mined cobalt came from artisanal mining in the DR Congo in recent years. Cobalt is a key material for electromobility, which is needed to produce rechargeable lithium-ion batteries. Together with the production of industrial mining, the Congolese cobalt production contributed to more than 60 per cent of the world market share.

**Platinum, chrome, titanium, nickel and vanadium deposits** are mined in the Bushveld Ultramafic Complex in Transvaal. The Paleoproterozoic Bushveld represents the world's largest layered intrusion and occupies the area between the Pilanesberg volcano in the west, Lydenburg in the east, Pretoria in the south, and Potgietersrus in the north (Republic of South Africa since 1994 divided into the provinces Northwest, Limpopo, Mpumalanga and Gauteng) and covers an area of 66,000 $km^2$ on which on 30,000 $km^2$ in addition to the mentioned deposits also iron ore deposits can be found. As the Bushveld magma cooled slowly, minerals of different compositions and crystallization temperatures formed and had time to form into crystals in the liquid magma or float to the top. This gave rise to the remarkably continuous layers of differing mineral composition of the Bushveld complex. It contains approximately 95 per cent of Africa's and 40 per cent of the world's platinum group metals production. Moreover, it contains nearly 90 per cent of the world's reserves. The Republic of South Africa is involved with around 20 per cent of the iron ore production of Africa and is also the world leader in the production of manganese and chrome iron ore.

Platinum group metals and chromite deposits come from the Great Dyke in Zimbabwe. It is estimated that the Paleoproterozoic, 480 km long intrusion contains a resource of over 4,500 million tons of platinum-bearing material. Chromium ore is also mined in Sierra Leone from serpentinized dunite, an intrusive igneous rock that dominantly consists of coarse grains of the pale green mineral olivine.

Gabon, like the Republic of South Africa and the DR Congo, is also considered one of the resource chambers of the industrialized countries. The mining sector is dominated by manganese and uranium mining. Manganese ore occurs on the residual Lower Proterozoic deposits of Mekambo, as well as gold and diamond deposits. Rich deposits of manganese ore are found at Moanda in the southeast of the country. Embedded within these are iron ore deposits. Deposits of uranium lie in the Mounana area, in the continental Franceville Basin. The Proterozoic Franceville Supergroup, consisting of sandstones, shales, and quartzites, covers the majority of the middle and upper Ogowe Basin in Gabon. Similar formations form the subsoil of a strip between the Mayumbe

Mountains. Uranium production was halted in 1999. However, reasonably assured resources of around 4,800 tons of uranium have not yet been exploited.

In the Guinea countries of West Africa, the **iron ore extraction** from hematite is tied to the highly metamorphic rocks of the Precambrian basement complex. Iron ore has been mined in Sierra Leone at Marampa-Lunsar since 1933. In the Simandou region in southeastern Guinea, the world's largest known extractable iron ore deposit was discovered in 1996. It is estimated to contain about 2.5 billion tons of ore and is not yet being exploited. However, there are actual plans for the commercial extraction of this high-quality ore, which has approximately 65 per cent iron content. In neighboring Liberia, iron ore has been mined since 1951 in the Bomi Hills and currently at the Mao River, at Mount Nimba, and in the Bong Hills. Hematite deposits have also been explored in the other states of the Guinea coast, but they are not currently being industrially exploited.

**Bauxite** for aluminum production occurs across Guinea. Guinea has about 50 per cent of the global bauxite reserves. The most significant deposit is located at Sangaredi, about 250 km north of Conakry. In 2020, 23 per cent of the world's bauxite production, or 86 million tons, took place there. Bauxite deposits in Ghana and Sierra Leone are exploited on a much smaller scale. Chromium deposits from the Kambu Hills in Sierra Leone, tin and tungsten deposits from the Jos Plateau in Nigeria, and manganese deposits from Grand-Lahou in the Republic of Côte d'Ivoire and Nsuta in Ghana are of only local and regional economic importance. More important for economic development are the readily exploitable coastal phosphate deposits found in Anècho, Togo.

In addition to oil and gas deposits in the Algerian and Libyan Sahara (see below), North Africa's most important mining products are phosphate deposits tied to Meso- and Cenozoic sediments. They are primarily depleted in Morocco, Mauritania, Tunisia, and Algeria. Smaller deposits are also exploited in West Africa in Togo and Senegal, as well as in South Africa. With around 10 per cent of the global market share, Algeria is one of the leading mercury producers. However, due to the high toxicity of mercury, which is now largely limited to scientific use, its extraction plays only a minor role.

**Coal deposits** are found in Africa in the Upper Carboniferous basins of Béchar in Algeria and Jerada in Morocco, in near-surface strata of Permian-Carboniferous age in South Africa and Zimbabwe, but also in Mozambique, Zambia, and Tanzania, as well as in the lower Eocene of Nigeria and the Congo Basin. In 2023, the Rep. of South Africa is the seventh-largest coal-producing country in the world and the fifth-largest coal exporter (2021). After gold, coal is South Africa's second-largest export.

The production of **hydrocarbons** (crude oil and gas) in Africa is limited to the North African countries of Algeria and Libya, and Western Central Africa (Nigeria, Chad, Niger, Gabon, Angola). In their order of importance, the productive reservoirs in Algeria are in Devonian and Carboniferous, Cambro-Ordovician and basal Triassic sandstones. The oil and gas fields lie in anticlines, faulted anticlines, and domes. The sealing rocks include Triassic to Lower Jurassic evaporitic sequences, Carboniferous and Devonian shales. The major source rocks were the widespread, organic-rich Silurian shales. In western Central Africa, the source rocks range from the Upper Cretaceous to the Neogene, also in the shelf area.

The first productive exploration drillings for oil in Algeria took place in 1956. Production in the Saharan oil fields of Eastern Algeria, Edscheleh and Hassi Messaoud, began in 1958. In 2023, oil production amounted to around 63.6 million tons, accounting for 1.4 per cent of global production (rank 17). In 2021, Algeria's crude oil reserves were estimated at 1.66 billion tons, which is more than 10 billion barrels, equivalent to an additional 27 years of production. In addition to crude oil reserves, Algeria has natural gas reserves of 2,279 billion cubic meters. The most important gas field is Hassi R'Mel in the Western Great Erg ("Grand Erg Occidental"). In 2021, just over 85.1 billion cubic meters were produced in Algeria, accounting for 2.2 per cent of global production (rank 10). The oil economy in Libya developed in a very similar way. The first successful drilling was carried out in 1957 in the Atschan field near the Algerian border. Until the first civil war in 2011, Libya, with an annual crude oil production of more than 77 million tons, was one of the ten largest oil-producing countries in the world. Since the new civil war in 2014 and the ongoing unrest, production had dropped to 20.6 million tons per year by 2021, accounting for only 0.4 per cent of global production, but in 2022 it again exceeded 51 million tons per year, moving Libya from rank 30 to rank 19 among the leading oil-producing countries. In 2024, the production increases up to 58 million tons.

In Central Africa, oil was first explored and produced in 1956 in the Niger Delta of Nigeria, and a little later in the offshore shelf area. Oil became the dominant factor in Nigeria's economic development and is still decisive for the OPEC state's supremacy in West and Central Africa. In 2024, Nigeria produced 95 million tons, accounting for 2.2 per cent of global production, ranking it 15th among the leading oil-producing countries. In contrast, deposits in the south of Benin, Ghana, and the Republic of Côte d'Ivoire are modest and only partially exploited. With 6.6 million tons, 0.2 per cent of global production, and thus ranking 38 of the producing countries in 2020, the production figures from the south of the Republic of Chad are still manageable. In Gabon, offshore oil production doubled from 1970 to 2024 to 13.4 million tons. Gabon thus ranks 36th among oil-producing countries with 2.7 per cent of global production and is a member of OPEC. The Atlantic or western sedimentary basin, which extends offshore along the coast at least as far as Sao Tomé, is the source of Gabon's coastal oil production. Trapping structures include faults related to the opening up of the Atlantic Ocean and salt diapirs. Gabon's largest oil field, onshore about 100 km north of Gamba, was discovered in 1985 and began production in 1989. With now (2024) 15.8 million tons and 3.3 per cent of global production (rank 33), the Republic of Congo has had higher production than Gabon since 2010 and has also been a member of OPEC since 2018. Angola, already a member of OPEC since 2007, has increased its production from 36.9 million tons in 2000 to 71.6 million tons in 2024. It is thus the second largest oil producer in Africa after Nigeria and ahead of Algeria, and ranks 18th in the global production statistics with 1.3 per cent (Table 3.1).

**Table 3.1** Natural mineral deposits in Africa (PRO = production 2021; RE = reserves 2021). Data from the U.S. Geological Survey: Mineral Commodity Summaries report from January 2022; 2021 data. a = Source: (2020) World Mineral Production: British Geological Survey; b = BGR—Federal Institute for Geosciences and Natural Resources (2022): BGR Energy Data 2021; c = unpublished estimate of the Federal Institute for Geosciences and Natural resources (BGR). (1) Usable ore, (2) crude ore, (3) palladium, platinum, iridium, osmium, rhodium, ruthenium, (4) only platinum, (5) ilmenite and rutile together, n. a.: not specified/not available/not found

| | # 1 | | # 2 | | # 3 | |
|---|---|---|---|---|---|---|
| **Bauxite**<br>*[1,000 Tons]* | Guinea | *Production (PRO): 123,000*<br>*Reserve (RE): 7,400,000* | Sierra Leone (a) | *PRO: 1,963*<br>*RE: N/a* | Ghana (a) | *PRO: 1,162*<br>*RE: N/a* |
| **Beryllium**<br>*[Tons]* | Mozambique | *PRO: 24*<br>*RE: N/a* | Madagascar | *PRO: 1*<br>*RE: N/a* | Uganda | *PRO: 1*<br>*RE: N/a* |
| **Chromite**<br>*[1,000 Tons]* | South Africa | *PRO: 18,435*<br>*RE: 200,000* | Zimbabwe | *PRO: 1,285 (a)*<br>*RE: 800 (c)* | Madagascar | *PRO: 12*<br>*RE: N/a* |
| **Coal**<br>*[1,000 Tons]* | South Africa (a) | *PRO: 229,000*<br>*RE: N/a* | Mozambique (a) | *PRO: 8,041*<br>*RE: N/a* | Zimbabwe (a) | *PRO: 2,751*<br>*RE: N/a* |
| **Cobalt**<br>*[Tons]* | D. R. Congo | *PRO: 220,000*<br>*RE: 3,500,000* | Madagascar | *PRO: 2,600*<br>*RE: 100,000* | Zambia | *PRO: 2,500*<br>*RE: 13,000* |
| **Copper**<br>*[1,000 Tons]* | D. R. Congo | *PRO: 5,300*<br>*RE: 31,000* | Zambia | *PRO: 680*<br>*RE: 21,000* | Morocco (a) | *PRO: 34*<br>*RE: N/a* |
| **Crude oil**<br>*[Million Tons]* | Nigeria (b) | *PRO: 95*<br>*RE: 5,019* | Angola (b) | *PRO: 71.6*<br>*RE: 1,050* | Algeria (b) | *PRO: 63.6*<br>*RE: 1,660* |
| **Diamonds** (industrial-grade)<br>*[Million Carats]* | D.R. Congo | *PRO: 8*<br>*RE: 150* | Botswana | *PRO: 7*<br>*RE: 300* | South Africa | *PRO: 4*<br>*RE: 120* |
| **Gold**<br>*[Tons]* | Ghana | *PRO: 130*<br>*Re: 1,000* | South Africa | *PRO: 100*<br>*RE: 5,000* | Mali | *PRO: 70*<br>*RE: N/a* |
| **Iron Ore**<br>*[1000 Tons]* | South Africa | *PRO: 61,000* [1]<br>*RE: 1,000,000* [2] | Mauretania | *PRO: 12,515 (a)*<br>*RE: 325,000 (c)* | Liberia | *PRO: 4,874 (a)*<br>*RE: 1,760,000 (c)* |
| **Lithium**<br>*[Tons]* | Zimbabwe | *PRO: 22,000*<br>*RE: 220,000* | Nigeria | *PRO: 130 (a)*<br>*RE: 30,000 (c)* | *no others* | |
| **Manganese**<br>*[1,000 Tons]* | South Africa | *PRO: 19,156*<br>*RE: 640,000* | Gabon | *PRO: 7,000*<br>*RE: 61,000* | Ghana | *PRO: 2,466*<br>*RE: 13,000* |
| **Natural gas**<br>*[Billion $m^3$]* | Algeria (b) | *PRO: 101.5*<br>*RE: 2,279* | Egypt (b) | *PRO: 62.1*<br>*RE: 2,139* | Nigeria (b) | *PRO: 50.0*<br>*RE: 5,848* |
| | **# 1** | | **# 2** | | **# 3** | |
| **Nickel**<br>*[Tons]* | South Africa | *PRO: 34,908 (a)*<br>*RE: 700,000 (c)* | Ivory Coast (a) | *PRO: 18,600*<br>*RE: N/a* | Zimbabwe (a) | *PRO: 16,336*<br>*RE: N/a* |
| **Phosphorites**<br>*[1,000 Tons]* | Morocco | *PRO: 38,000*<br>*RE: 50,000,000* | Egypt | *PRO: 5,000*<br>*RE: 2,800,000* | Tunisia | *PRO: 3,200*<br>*RE: 100,000* |
| **Platinum group metals** [3]<br>*[Kilograms]* | South Africa | *PRO: 125,000*<br>*RE: 63,000,000* | Zimbabwe | *PRO: 19,200*<br>*RE: 1,200,000* | Ethiopia [4] (a) | *PRO: 2*<br>*RE: N/a* |
| **Rare Earths**<br>*[1,000 Tons]* | Tanzania | *PRO: 890*<br>*RE: N/a* | Madagascar | *PRO: 3.2*<br>*RE: N/a* | Burundi | *PRO: 0.1*<br>*RE: N/a* |
| **Tantalum**<br>*[Tons]* | D.R. Congo | *PRO: 700*<br>*RE: N/a* | Rwanda | *PRO: 270*<br>*RE: N/a* | Nigeria | *PRO: 260*<br>*RE: N/a* |
| **Tin**<br>*[Tons]* | D. R. Congo | *PRO: 19,000*<br>*RE: 130,000* | Nigeria | *PRO: 8,100*<br>*RE: N/a* | Rwanda | *PRO: 3,800*<br>*RE: N/a* |
| **Titanium Minerals** [5]<br>*[1,000 Tons]* | South Africa | *PRO: 1,090*<br>*RE: 36,500* | Mozambique | *PRO: 979*<br>*RE: 26,890* | Senegal | *PRO: 370*<br>*RE: N/a* |
| **Tungsten**<br>*[Tons]* | Rwanda | *PRO: 950*<br>*RE: N/a* | Burundi (a) | *PRO: 165*<br>*RE: N/a* | D. R. Congo (a) | *PRO: 128*<br>*RE: N/a* |
| **Uran**<br>*[Tons]* | Namibia | *PRO: 5,613*<br>*RE: 279,400 (b)* | Niger | *PRO: 2,020*<br>*RE: 238,700 (b)* | South Africa | *PRO: 200*<br>*RE: 236,600 (b)* |
| **Vanadium**<br>*[Tons]* | South Africa | *PRO: 9,100*<br>*RE: 3,500,000* | *no others* | | *no others* | |
| **Vermiculite**<br>*[1,000 Tons]* | South Africa | *PRO: 140*<br>*RE: 14,000* | Zimbabwe | *PRO: 30*<br>*RE: N/a* | Uganda | *PRO: 10*<br>*RE: N/a* |
| **Zinc**<br>*[Tons]* | South Africa | *PRO: 160,816 (a)*<br>*RE: 8,000,000 (c)* | Eritrea | *PRO: 121,930 (a)*<br>*RE: 1,800,000 (c)* | Namibia | *PRO: 62,333 (a)*<br>*RE: 1,000,000 (c)* |
| **Zircon**<br>*[Tons]* | South Africa | *PRO: 270*<br>*RE: 5,900* | Mozambique | *PRO: 110*<br>*RE: 1,800* | Senegal | *PRO: 70*<br>*RE: N/a* |

## References


Baumhauer R (2024) Physische Geographie—Geomorphologie, 4th edn. Borntraeger, Stuttgart, p 180

Brooks JJ, Jarrett A, Sirantoine E (2016) Early sponges and toxic protists: possible sources of cryostane, an age diagnostic biomarker antedating Sturtian Snowball Earth. Geobiology 14:129–149

Catuneau H, Wopfner PG, Erikson BS, Rubidge R, Smith MH, Hancox PJ (2005) The Karoo basins of south-central Africa. J Afr Earth Sci 43:211–253

Hoffmann PF, Schrag DP (2002) The snowball earth hypothesis: testing the limits of global change. Terra Nova 14:129–155

Hoffmann PF, Kaufman AJ, Halverson GP, Schrag DP (1998) Neoproterozoic snowball earth. Science 281:1342–1346

Key RM (1992) An introduction to the crystalline basement of Africa. In: Wright EP, Burgess WG (eds) Geol Soc Spec Publ 66:29–57

Kirschvink JL (1992) Late Proterozoic low-latitude glaciation: the snowball earth. In: Schopf JW, Klein C (eds) The Proterozoic biosphere. Cambridge University Press, pp 51–52

Klimm E (1985) Die Guineaänder. In: Klimm E (ed) Afrika 2. List, Munich, pp 60–109

Klitsch E (1970) Die Strukturgeschichte der Zentralsahara. Geol Rdsch 59:459–527

MacGabhann BA (2005) Age constraints on Precambrian glaciations and the subdivision of Neoproterozoic time. IUGS Ediacaran Subcommission Circular, August 21, p 14

Manshard W (1981) Afrika—südlich der Sahara. Fischer, Frankfurt, p 328

Nyamweru CK (1996) The African rift system. In: Adams WM, Goudie AS, Orme AR (eds) The physical geography of Africa. Oxford University Press, New York, pp 18–33

Petters SW (1991) Regional geology of Africa. Springer, Berlin Heidelberg, p 695

Schamp EW (1985) Mittelafrika. In: Klimm E (ed) Afrika 2. List, Munich, pp 110–151

Schlüter T (2008) Geological atlas of Africa. With notes on stratigraphy, tectonics, economic geology, geohazards, geosites and geoscientific education of each county, 2nd edn. Springer, Berlin Heidelberg, p 307

Summerfield MA (1996) Tectonics, geology, and long-term landscape development. In: Adams WM, Goudie AS, Orme AR (eds) The physical geography of Africa. Oxford University Press, New York, pp 1–17


## Further Readings


Baumhauer R, Kneisel C, Möller S, Schütt B, Tressel E (2017) Einführung in die Physische Geographie, 3rd edn. WBG, Darmstadt, p 352

Dawson JB (2008) The Gregory rift valley and Neogene-recent volcanoes of northern Tanzania. Geological Society Memoir No. 33. Geological Society of London

De Wit MJGF, de Wit MCJ (2015) (eds) Geology and Resource Potential of the Congo Basin. Springer, Berlin, Heidelberg, p 417

Idoine NE, Raycraft ER, Shaw RA, Hobbs SF, Deady EA, Everett P, Evans EJ, Mills AJ (2022) World mineral production 2016–20. British Geological Survey, Keyworth, Nottingham, p 98

Muwanga A, Owor M, Schumann A, Kawule W (2007) (eds) International conference on the East African rift system, Abstract vol 1, Kampala, Uganda, pp 1–94

Rad VU, Hinz K, Sarnthein M (eds) (2012) Geology of the Northwest African continental margin. Springer, Berlin, Heidelberg, p 711

Schandelmeier H, Thorweihe U (eds) (2017) Geoscientific research in Northeast Africa. CRC Press, Bota Raton, p 793

Tawadros E (2011) Geology of North Africa. CRC Press, Boca Raton, p 930

U.S. Geological Survey (2022) Mineral commodity summaries. https://doi.org/10.3133/mcs2022

Varet J (2017) Geology of Afar (East Africa). Springer, Berlin, Heidelberg, p 336

Viljoen R, Viljoen MJ, Anhaeusser CR (eds) (2016) Africa's top geological sites. In: 35th international geological congress commemorative volume. Struik Nature, Cape Town, South Africa

# 4 First-Order Landforms

Within the African continent, the first-order morphological distinction is between the broadly elevated east and south, and the lower-lying north and west. Large parts of Northern and Western Africa lie between 400 and 600 m (Low Africa). Southern and Eastern Africa are characterized by average altitudes between 800 and 1,000 m (High Africa). The boundary is formed by the 1,000 m contour line. It extends from the Ethiopian Highlands via the western slope of the Central African Ridge and the northern slope of the Lunda Ridge to the highlands of Bié in central Angola. The land area proportions at the various altitude levels are: <200 m above sea level (a.s.l.) 10 per cent, 200–1,000 m a.s.l. 63.5 per cent, 1,000–2,000 m a.s.l. 23 per cent, >2,000 m a.s.l. 3.5 per cent (Figs. 4.1 and 4.2).

The analysis of the hypsometric data shows that 80 per cent of Northern Africa (north of 5° N latitude) lies below the mean global elevation of the continents, while 70 per cent of Southern Africa (south of 5° N latitude) is above the global mean. Characteristic of the African relief is the absence of both mountains formed by folding and alpine mountains above 4,000 m as well as widespread lowlands (Fig. 4.3).

The outline shape of the African continent reflects the uniformity of a slightly structured coastline. The coastal environments are geomorphologically hardly investigated. Because of Late Mesozoic and Cenozoic tectonic uplift along the ridges, and the Atlas Mountains and South African Cape Fold Belt in the northwest and extreme south, larger coastal plains with a simplified coast are found exclusively on the east coast from Mozambique to Somalia and the Namib coast. More than 20 per cent of the African coast is characterized by steep coasts. The monotony of the coasts is the result of a long geological and morphological development, in which old mountain ranges have been leveled and the continent had no orogenies for several hundred million years (Chap. 3). Among the long-lasting, widespread tectonic movements, new rock masses have been eroded from the resulting ridges and deposited in nearby basins. Faults of several hundred meters of vertical separation were leveled by erosion processes, and fold structures were cut. Until the Middle Tertiary, the morphology of the African continent showed a low relative altitude, little relief intensity, and a thick cover of heavily weathered rock, the **African (Erosion) Surface**. The African Surface is the oldest and most extensive erosion surface of the African continent. It is of polygenetic origin and has developed over a period of 10 to 100 Ma. Since the Oligocene The African Erosion Surface has been divided



R. Baumhauer, *Physical Geography of Africa*, Geography of the Physical Environment,
https://doi.org/10.1007/978-3-662-71406-5_4

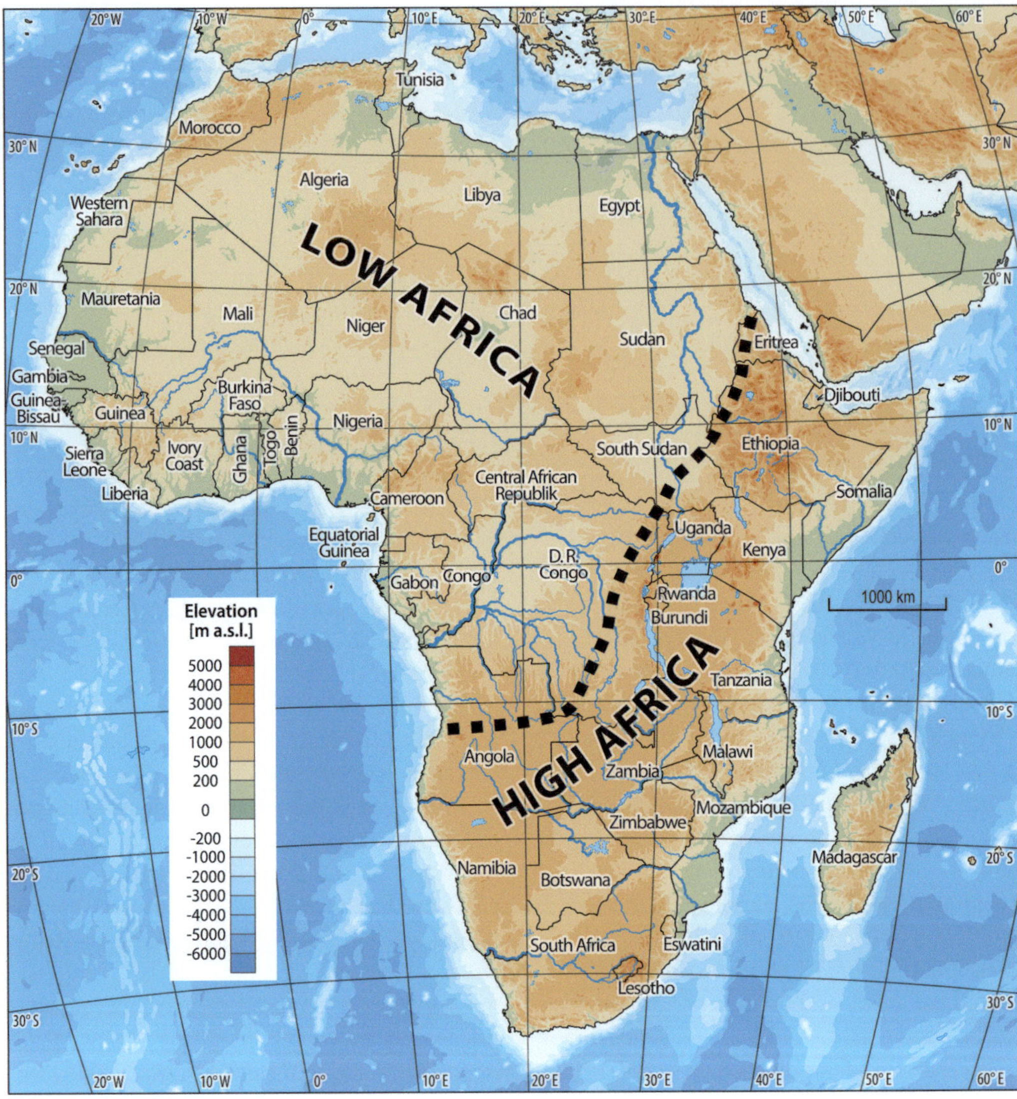


**Fig. 4.1** Africa physical map with low and high Africa boundary

into ridge areas and sediment basins by uplift movements associated with the activation of the **Afar, Karoo, and Walvis (Super-) Plumes** (see also Fig. 3.8). The position of these **ridges and basins** determines the current large-scale morphology of Africa. Up to date, the shallow burial, even from the ridge areas, is largely eroded and only still present where it was protected from erosion by partly thick **iron pans** (Petroplinthite according to WRB, formerly: **Laterite**) or bauxite pans. The ridges, which have a mean height of 1,500 ± 500 m, are always highest where they were covered by volcanic flood basalts, as in East Africa and central West Africa. The **Ethiopian Ridge** rises steeply from sea level to 2,000–2,400 m

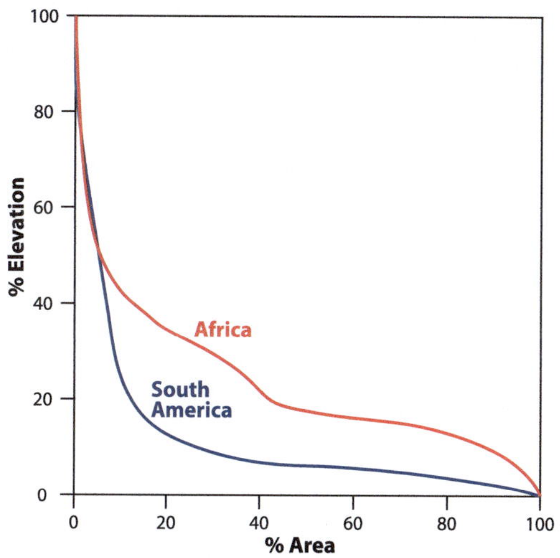


**Fig. 4.2** Comparison of hypsometric curves constructed from one-minute digital elevation data for Africa and South America

and is part of the **East African Ridge,** which extends via Kenya (3,000–4,000 m) with some interruptions to South Africa. The **East African Rift** has interrupted this ridge and thus created a unique block-faulted area of horsts and grabens, which is linked to strong volcanic activity with volcanic massifs and extensive lava flows. Other examples of volcanism in ridge areas are Kilimanjaro and Mt. Kenya, but also the Hoggar Mountains and the Tibesti in the Sahara and the Darfur Mountains in Sudan. Due to erosion, some ridges are not topographically visible, for example, the Senegal Ridge is hardly recognizable in the relief.

In southern Africa, the horseshoe-shaped, more than 1,000 m high and 5,000 km long **Great Escarpment** separates the narrow coastal plain from a high inland plateau. The Great Escarpment is also the boundary of the **African (Erosion) Surface**, in which almost all geological sequences of southern Africa are represented. The Great Escarpment begins in northwest Angola, crosses Namibia and South Africa, Lesotho, and the Kingdom of Eswatini, and extends to eastern Zimbabwe and Mozambique. In southern Angola, the elevation of the escarpment is about 2,600 m, and the Drakensberg in KwaZulu-Natal culminates up to 3,400 m. However, the escarpment is maturely dissected or, as in Central Namibia, even completely dissolved. The escarpment occurs after the breakup of Gondwana in the Late Jurassic and Early Cretaceous, within the last 30 Ma. It is typical for passive continental margins. It thus represents a passive eroded residue of a continental boundary that has retreated from its original position, and thus also explains the relatively coast-parallel run of the Great Escarpment 50–200 km from the coastline. Its morphogenesis is very complex, and although there are differences between the drier western part, and the wetter eastern part, morphologically the Great Escarpment is one unit.

In addition to the ridges, large basins enclosed by the higher ridges are typical for the African continent. The most important are the Sudan- (White Nile-) Basin, Congo Basin, Chad-, El-Djouf- (Western Sahara- or Niger-) and the **Kalahari Basin**. The **Niger Basin and** the **Congo Basin** have large exorheic river systems (Niger and Congo), and the **Chad Basin** drains endorheically. The sediment cover is relatively shallow both in the three large intracontinental basins, the Kalahari and Chad Basin, and the Congo Basin**,** as well as in the many smaller basins of the African continent. Since the uplift of the ridge areas began in the Jurassic and continued to intensify in the Oligocene and Miocene, non-marine sediments of aeolian, limnic, fluvial, and colluvial origin have been deposited in the basins. In the Chad Basin, there is a maximum of 1,000 m of sedimentary rock younger than 30 Ma. They superpose the African Surface. In the Kalahari Basin, the Kalahari Group covers the **African (Erosion) Surface**, which is less than 100 m thick and reaches a maximum thickness of 400 m (Fig. 4.4).

The **Kalahari Basin** is the largest in High Africa. Its lowest point is in the Makgadikgadi salt pan in Botswana at an altitude of about 900 m

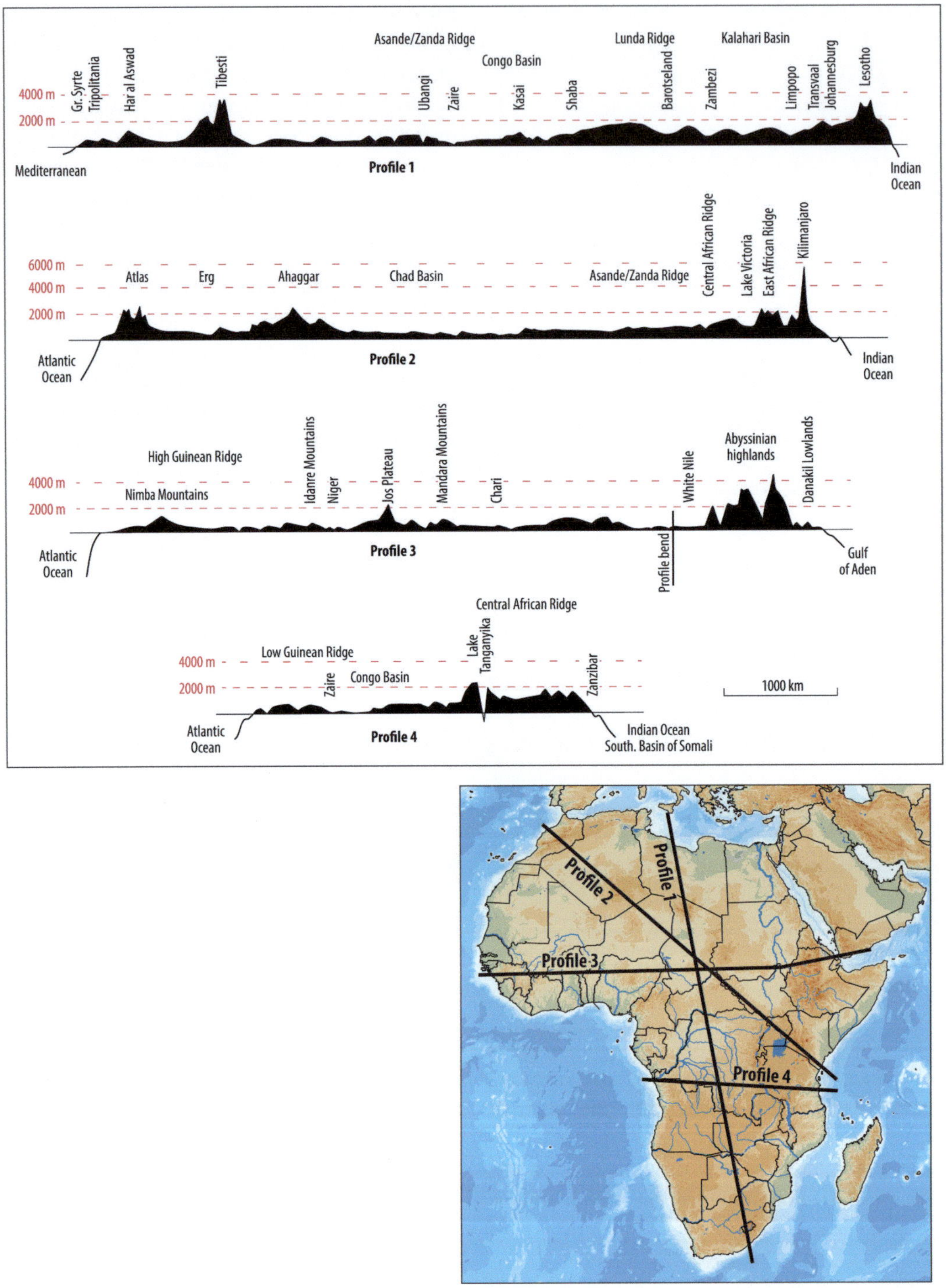


**Fig. 4.3** Topographic profiles across Africa. Modified from Manshard (1981)

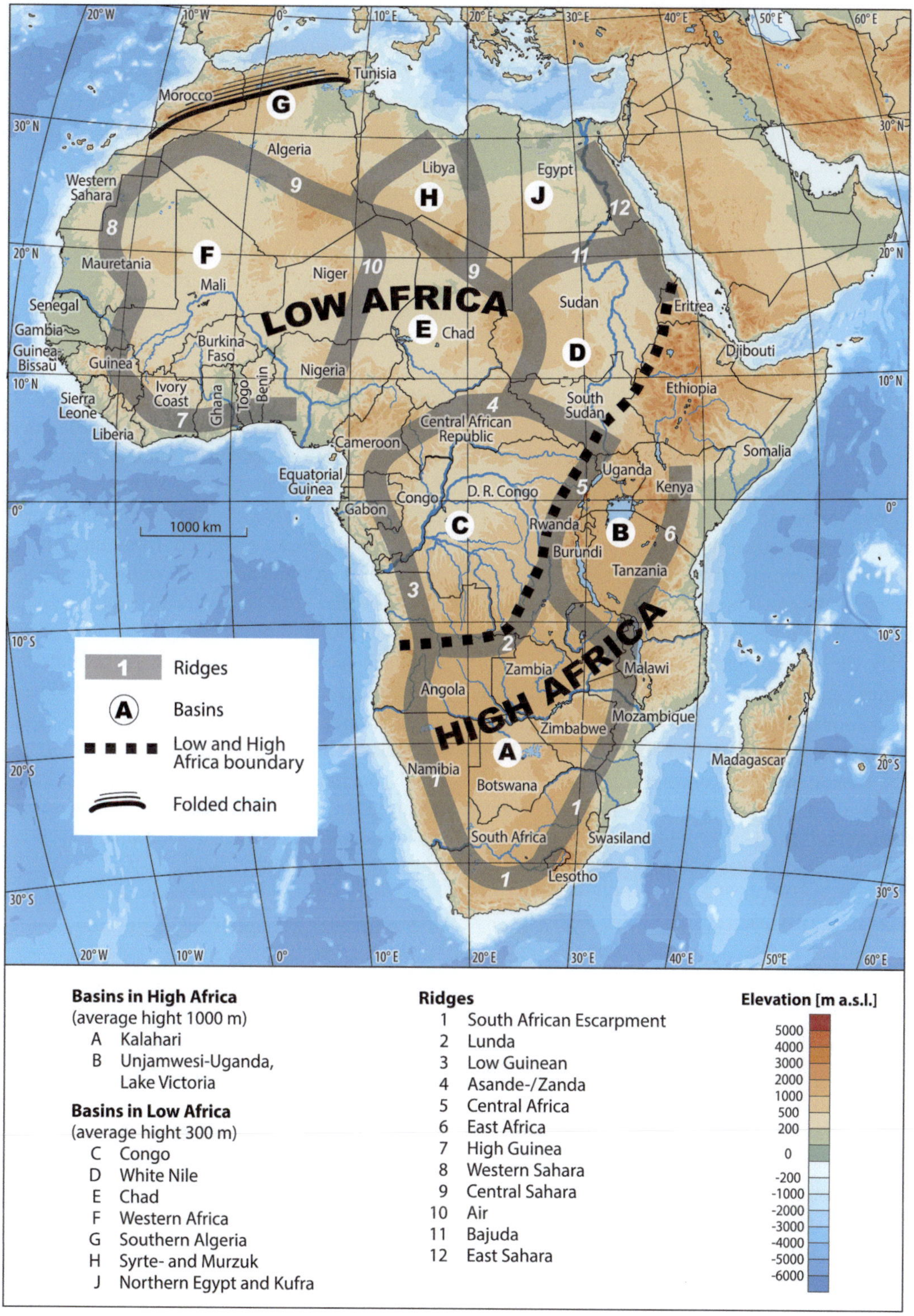


**Fig. 4.4** The geological ridges and basins of Africa

a.s.l. The surrounding ridges of southern Africa have elevations of 2,000–2,500 m and rise in the Drakensberg to almost 3,500 m. To the north, the South (east) African Ridge transitions into the Njassa Ridge, which forks into the Central and East African Ridge. Both enclose the Unjamwesi Basin with Lake Victoria, whose water level is at 1,134 m. It has an average depth of 40 m and a maximum depth of 85 m. The lakes in the neighboring graben basins are considerably deeper. For example, Lake Tanganyika has a maximum depth of 1,435 m. The Turkana Basin separates the Ethiopian Highlands from the **East African Ridge**. The highlands, with extensive high plateaus at 2,000 m and summit meadows exceeding 4,000 m, are the largest elevations of the African continent.

Within the African continent itself, the first-order morphological distinction of Lower Africa is less clear because the ridges are hardly visible in the landscape. The most distinct is the Congo Basin, which has an elevation of 300–500 m. Only the eastern parts at the border to High Africa, the **Central African Ridge,** and in the Cameroon Highlands, the northern section of the **Lower Guinea Ridge** lie above 2,000 m. The **Lunda Ridge** in the south and the **Asande Ridge** in the north, also known as the **North Equatorial Ridge**, are widespread. The Sudan includes the (White) Upper Nile—and the Chad Basin—both separated by the **Darfur Ridge**—and the Niger Basin. The West Sudan is bounded in the south by the Upper Guinea Ridge, which lies up to 1,500 m only in some landscape elevations. The Sahara is diagonally divided by the **Middle Saharan Ridge** occupied by the mountains of Hoggar and Tibesti. The only relatively closed basin occurred in the **Qued Irharrhar ("Igharghar Basin")** of the Great Eastern Erg, while the Libyan Basin and the western Sahara are subdivided into sub-basins by smaller ridges.

Apart from the continental margins, the **African (Erosion) Surface** usually has thicknesses of sediments less than 100 m. Following the Oligocene uplift, the erosion processes in the Miocene and Pliocene formed two **Post-African Surfaces**. Where granites are exposed, inselberg landscapes have been created by the uplift and subsequent erosion. The first-order morphology of the Precambrian basement complex is a **Peneplain-Inselberg-Relief** (see Excursus).

Due to its narrow shelf area, the African continent has only a few islands. The Atlantic islands have an ancient crystalline base, indicating genetic similarities with the continent, overlain by a younger volcanic surface. In contrast, Indian Ocean islands are either volcanic, such as the Comoros and Mascarenes, or granitic uplifts of submarine ridges, which are partly occupied with coral structures. Examples of the latter include the Seychelles and Sokotra, an offshoot of the Somali Peninsula. The islands of Lamu, Zanzibar, Pemba, and Mafia Island off the East African coast have a core of coral limestone and a young Pliocene or Pleistocene sediment cover. Madagascar separated from the African continent with India about 150 million years ago, and separated from India in the Late Mesozoic era about 90 million years ago. Its eastern part is a fault-block mountain range of granites and gneisses, which is dominated by young volcanoes among which the Tsaratanana with 2,876 m and the Ankaratra massif with 2,643 m are the highest points. In the narrow western part of the island occurred an escarpment landscape of Mesozoic and Cenozoic limestone and sandstones.

**Excursus: Peneplains, Pediplains and Inselbergs**

Broad plains, often with isolated hills (inselbergs) or mountainous, cover large areas of tropical zones. The plains are of two types: erosional and sedimentary. Erosional plains develop on variable lithologies and structures, whereas sedimentary plains are zones where alluvial and lacustrine sediments accumulate. Davis (1899) proposed the term peneplain for plains in humid areas and King (1942) proposed pediplain for arid zones. These plains are part of the present-day landscape. Peneplains are rounded and undulating plains that exhibit very little elevation contrast, and their surface shape is not influenced by differences in rock or geological structures. Instead, peneplains cut across different geological structures and rocks, thereby forming extensive erosion surfaces. The formation of peneplains in Africa on the cratons or shields suggests that they had a very long period of tectonic calm available for their genesis. If parts of peneplains are lifted due to tectonic movements, a peneplain staircase with intervening peneplain steps (German: "Rumpfstufe", French: "Falaise") can form. Ahnert (2015) shows this in Kenya with the basement uplift in three steps from the Indian Ocean to the highlands around Nairobi. The three denudatively formed surfaces at 500 m a.s.l., 1,000 m a.s.l., and 1,500 m a.s.l. represent uplifted peneplains that form a peneplain staircase with steps separating the three surfaces. Therefore, genetically, the three peneplains occurred by tectonic uplift processes, between which very long-lasting tectonic repose periods resulted in extensive denudation.

Inselbergs (or "bornhardts", domed inselbergs after the geologist Wilhelm Bornhardt, as a mesoform also "tor", boulder inselberg) describe steep rocky hills that stand out over the (tropical) plains. Inselbergs develop primarily in tropical and subtropical cratonic regions, preferably on Precambrian granites, gneisses and/or migmatites. Inselbergs reach diameters of 10 m to several hundred meters and elevations of 1–2 m (shield inselberg) to several hundred meters. They are formed by self-energizing processes during deep weathering and erosion of exhumed bedrock and its surroundings, saprolitization, slope erosion, and stream cut. They pass over with a noticeable change to the more gently sloping surface, which is the proposed "pediment". Inselbergs are most impressively developed in the wet-dry tropical climate (wet Savanna) areas. Zonal inselbergs are present in a peneplain scarp (e.g., during the Great Escarpment in South Africa), and azonal inselbergs are linked to the change of the drainage network (Fig. 4.5b). Hard rock Inselbergs occur by the exhumation of morphologically more resistant rock, most abundantly granites (Figs. 4.5a and 4.6).

## 4.1 The Geomorphology of the Maghreb and the Mediterranean Region of Africa

The Maghreb countries of North Africa (Morocco, Algeria, Tunisia; Maghreb or Maghrib means the western region of the Arab world) are dominated by the 220–250 km wide **Atlas Mountains**, which, as an independent geomorphological first-order unit of the African relief, is strongly structured in itself. They extend for more than 2,000 km, from the Moroccan port of Agadir in the southwest to the Tunisian capital of Tunis in the northeast. A northern mountain range, consisting of **Rif and Tell Atlas**, and a southern one with **High Atlas**, and **Sahara Atlas** form the outer framework. The central section is dominated by the Moroccan Meseta, also referred to as the **Highlands of the Chotts**, an intramontanous high plain. Both are separated by the **Middle Atlas**. In the eastern part of the Tunisian **Dorsale**, the alpine mountain area transitions into lower mountain ranges. The approximately 2,100 km long southern boundary to the Sahara is marked by the Wadi Draa, which runs south of the **Anti-Atlas** and drains to the Atlantic. In the northwest, the Rif Mountains approach the Strait of Gibraltar up to 14 km from the Iberian Peninsula (Fig. 4.7).

The present mountains were mostly formed during the Alpine orogeny when profound folding and faulting of sediments of the Tethys occurred in several phases during the

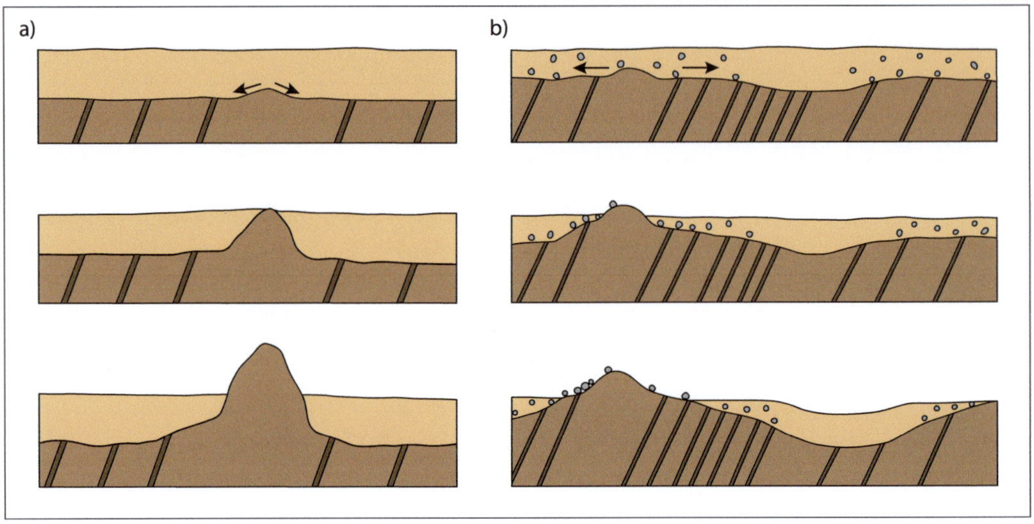


**Fig. 4.5** Development of inselberg landforms according to Bremer and Sander (2000, p. 12). Inselbergs arise from a thick weathering deposit. The initial stages of a weathering front can be attributed to the structure of the rock, the location of a later developing steep slope or a joint (**a**). As soon as the soil water movement is redirected, a feedback process occurs. If the chemical alteration mantle is differently deeply weathered (**b**), blocks can be excluded from weathering and cover the inselberg after exhumation, according to a semi-humid climate, more intense weathering, or a special lithology. All other factors being equal, climate is the decisive factor. Since water can also penetrate vertically through joints in granite or migmatite, inselbergs form particularly easily in these rocks, often with convex slopes

**Fig. 4.6** Granite peneplain and shield inselbergs, Northern Nigeria. *Photo* D. Busche

**Fig. 4.7** Anti-Atlas mountains with Wadi Draa at the height of Ouarzazate, Southern Morocco. *Photo* M. Lange, Mauritius Image

collision of the African with the Eurasian plate. Analogous to the folding of the Alps, the Pyrenees, and the Betic Cordillera on the Iberian Peninsula, marine sediments from the Tethys Sea have been compressed, folded, and lifted since the Cretaceous period. A process that continues due to the further advance of the African against the Eurasian Plate and repeatedly leads to earthquakes, for example, in September 2023 with a magnitude of 6.8 in the High Atlas of Morocco with almost 3,000 deaths or in 1980 with a magnitude of 7.5 near El Asnam, 150 km southwest of Algiers with more than 3,000 deaths. The oldest fold belt is the Rif Atlas, a coastal range running south-eastwards from northern Morocco and continuing into Algeria as the Tell Atlas. The High Atlas extends north-eastwards from the Atlantic coast north of Agadir into Algeria, where it becomes the Saharan Atlas. In eastern Algeria, the Saharan Atlas the Saharan and the Tell Atlas approach each other. The Middle Atlas is a plateau region in Morocco bounded by the high peaks of the High Atlas to the south. The Anti-Atlas marks the edge of the African Precambrian basement. At its eastern extreme, it converges with the High Atlas at Jebel Siroua, a volcanic outcrop. The Moroccan Anti-Atlas and parts of the Sahara Atlas are older and formed during the Variscan orogeny in the younger Paleozoic. Between their folded chains are also isolated remnants of the Precambrian basement.

The highest elevation is Mount Toubkal (4,165 m) in the High Atlas in Morocco. The different mountain ranges are separated from one another by deep endorheic basins and widespread peneplains, the High Plains ("Hautes Plaines"). The Moroccan Meseta forms a maturely dissected plain from younger sedimentary rocks. It slopes down to the Cenozoic Atlantic coastal plain. Further south to the Senegal delta, a flat, Saharan-influenced simplification coast follows. The 1,700 km long Mediterranean coastal section of the Maghreb Atlas countries extends from North Tunisia over Algeria to East Morocco and is designed as a steep coast (mountain longitudinal coast). As a result of the Alpine tectonics, the coast has cliffs that are repeatedly interrupted by small bays, while on the Tunisian east coast at the Gulf of Hammamet and the Gulf of Gabes, wide coastal plains predominate.

The coastal plains continue to the east over the Dschifara, the most important agricultural area of Libya, and the eastern half of Libya, including Cyrenaica. In northwest Libya, the former province of Tripolitania, the coastal plain ends in the south at the edge of Jebel Nefusa (728 m), a cuesta of Late Mesozoic Nubian sandstone and Cenozoic limestones. The Coastal lowland of the Great Syrte, on the other hand, extends further south. In the western border area, the hinterland of the Great Syrte, the volcanic area of Harudsch Al-Aswad rises with lava domes and lava sheets. In Egypt, Cenozoic limestone cuestas and scarped tablelands (mesas) dominate in the northern part bordering Libya. In the southern part, they consist of Nubian Sandstone as well. These limestone and sandstone cuestas and plateaus of the Libyan Desert extend via the Nile Valley to the east. The Atbai, fault-block mountains made of crystalline shales, old eruptive and intrusive masses, are heavily eroded, and their western foothills are heavily dissected. To the south, it extends to the peneplains of the Eritrean Highland.

The Cataracts of the Nile (**Nubian Cataracts**) are whitewater rapids of the Nile River between Khartoum and Aswan, where the surface of the water is broken by many small boulders and stones jutting out of the riverbed, as well as many rocky islets. The First Cataract is located just south of Aswan. It was selected for the construction of Aswan Low Dam, the first dam built across the Nile. The Second Cataract (or Great Cataract) is in Nubia (Sudan) and is now submerged under Lake Nasser. It is located 10 km south of the former site of Wadi Halfa, at the current location of the town. The Third Cataract is at Tombos/Hannek. The Fourth Cataract is in the Manasir Desert, and since 2008, is submerged under the reservoir of Merowe Dam. The Fifth Cataract is between Abu Hamad and Atbara, Sudan, and the Sixth Cataract is where the Nile cuts through the Sabaluka pluton, close to Bagrawiyah. With the sixth cataract, the Nile Valley begins somewhat north of Khartoum (376 m). The second cataract and the first cataract at Aswan are excavated into the Precambrian (crystalline) basement complex. The Nubian Ridge has diverted the river's course to the west while keeping its depth shallow and causing the formation of cataracts. The valley itself is embedded in the Nubian Sandstone. Fluvial gravel in the area of the valley slopes, the valley shoulders, and some branching dry valleys indicates changes in the course of the Nile. The Nile Valley itself is, on average, 11 km wide, the narrowest point in the section of the Nubian cataracts at Jebel as-Silsila 60 km north of Aswan measures only 350 m. At the widest point at Beni Suef south of Cairo, a low sill (ridge, swell) of Eocene limestones separates the 23 km wide Nile Valley from the El-Faijum (Fayoum) Basin to the west. The **Nile Delta** is formed in Lower Egypt, where the Nile River spreads out and drains into the Mediterranean Sea. It is an arcuate delta, as it resembles a triangle or flower when seen from above. The Nile Delta is one of the world's larger deltas and begins slightly downriver from Cairo. From Alexandria in the west to Port Said in the east; it covers 240 km of the Mediterranean coastline and is a rich agricultural region. From north to south, the delta is approximately 160 km in length. The delta covers an area of around 26,000 $km^2$ and is the most densely populated region of Africa. West of the Nile Delta, a steep coast with cliffs and narrow terraces is followed by a flat equilibrium coast about 3,000 km long up to South Tunisia, characterized by Cenozoic sediments and small cliffs, beach ridges, lagoons, and dunes.

The Atlantic coast in Morocco is characterized by 100–600 m high cliffs, flat terraces, and up to 80 m high dunes on beaches. To the south of the Senegal delta, a flat, Sahara-influenced simplified coast follows. South of the Senegal delta, the dominant steep coasts on the Atlantic are made of Precambrian basement and Paleozoic sediments, alternating with flat coastal sections lined with mangroves.

**Excursus: Cataract and Delta**

The term "cataract" in its original Greek meaning comes from the words "to flow down", although the original Greek term was the plural only and refers to a succession of waterfalls. The term waterfall is applied where there is a vertical fall of water. Rapids are less marked irregularities in the channel bed of a river, consisting of short, steep reaches that may be entirely submerged at a high stage. The cataracts of the Nile refer to flat, rugged stretches of the Nile between Aswan in Egypt and Khartoum in Sudan, where Granite barriers are broken through by the Nile. There are six cataracts, of which the first lies in present-day Egypt and the remaining five in Sudan. In earlier centuries, these six areas of the Nile formed the natural boundaries between different kingdoms.

Deltas are protuberances extending out from shorelines formed when rivers enter the ocean, seas or lakes, and supply sediment more rapidly than they can be redistributed by coastal processes. The Nile Delta is an "arcuate" delta (arc-shaped), the most fully developed equilibrium form of large deltas, which always arises when individual river (delta) arms have advanced so far into the (Mediterranean) sea that the ocean currents no longer allow the deposition of the load of the river (e.g., Nile) and no progradation occur, but the transported sediments are deposited offset sideways by the ocean currents.

## 4.2 The Geomorphology of Low Africa

The **Sahara** is the world's largest desert region, which stretches from the Atlantic coast to the Red Sea over 5,500 km and from the Mediterranean to the Niger River and the margins of the Ethiopian Highlands. The desert covers an area of approximately seven million square kilometers, as large as a quarter of the African continent and large parts of Lower Africa. The northern boundary of the Sahara lies on the southern piedmont of the Atlas Mountains. It is marked by the fault and fracture line ("Sahara line") running from Gabès in Tunisia, along the southern edge of the Atlas Mountains to Agadir on the Moroccan Atlantic coast.

Its southern boundary is much less distinct, and the desert grades into the semi-arid Sahel zone around 15° N or the latitude of the Niger bend and Lake Chad. The Sahara is characterized by generally low relief. The only major mountain areas occur in the central parts of the desert and include the volcanic mountains of the Tibesti (Emi Koussi, 3,415 m), Hoggar (Tahat, 2,918 m), Air Mountains, and Ennedi Mountains, which reach elevations of as much as 3,300 m. On the other hand, the Qattara Depression, lying below sea level (−133 m), is the deepest point of Low Africa. Elsewhere, there are extensive plateaus developed on flat-lying or gently tilted Paleozoic and Mesozoic sedimentary rocks and bounded by prominent cuestas. The surfaces of the plateaus consist either of bare rock with coarse blocks (Hammada) or gravel (Serir/Reg). Intervening basins developed after the Cenozoic uplift of the Central Sahara, as well as by differential erosion of sedimentary rocks. Many of these basins are occupied by extensive sand seas. They are also **Ergs or Edeyen** called which cover about 28 per cent of the area of the Sahara. A large part of the dune areas, especially in the southern Sahara, are of prehistoric origin and today stationary (see Chap. 9). **Wadis** or Ennedis (dry valleys) are elements of fluvial shaping, important components of the desert relief. Ephemeral drainages are collected in endorheic basins, often occupied by salt flats or salt marshes, **Sebhkas** or **Chotts**. Within the Sahara itself, a first-order geological and morphological distinction is between a western, a central, and an eastern section.

In the **Western Sahara**, the West Saharan Ridge is shaped by the peneplain of Karet-Yetti, to which the vast flat-lying rocks and cuestas of the basins of Tindouf and Taoudenni connect. They underlie the limnic and aeolian formations of the El-Djouf Basin in the southeast and those of the Ouahila Basin in the northeast. In the west, flat-lying Cenozoic sandstones and limestones extend to the coast. The valley of the Wadi Draa in southern Morocco forms the northern boundary of the Sahara with the Anti-Atlas.

The **Central Sahara** is dominated by the mountain areas of the Hoggar and the Tibesti. The extensive Hoggar Mountains are a fragmented uplift of the Precambrian basement complex, which is occupied by young volcanoes. Individual blocks are strongly uplifted, so the Atakor massif (2,500 m). The highest elevations are reached by the volcanic ruins of Ilamane (2,760 m) and the basaltic sheet of Mont Tahat (2,918 m). The Paleozoic sedimentary rocks surrounding the Hoggar Mountains form a ring of cuestas (the Tassili) and plateaus. Intervening alluvial sand and salt clay plains developed in the erosion basins. A southwestern off-shoot of the Hoggar Mountains is the Adrar des Iforas and the southern Aïr Mountains. Both are also flat uplifts of the basement with peneplains and inselbergs. In contrast to the Aïr Mountains, the Adrar des Iforas are without young volcanism. To the east, the basement extends to the Tibesti, which is also a fault block mountain range, but is largely made of sandstone. Young volcanism has created powerful basaltic sheets and volcanic cones such as the Emi Koussi (3,415 m), the highest mountain in the Sahara (Figs. 4.8 and 4.9).

Northeast of the Tibesti Mountains extends the Libyan Basin as the largest part of the **Eastern Sahara**. Gilf Al Kabir area (1,083 m) in southwest Egypt and the massif of Jebel Uwainat are developed on flat-lying or gently tilted Late Mesozoic Nubian Sandstone and Cenozoic limestones and bounded by prominent cuestas and mesas (scarped tablelands). In the western border area, in the hinterland of the Great Sirte, the volcanic area of Harudsch Al Aswad rises with lava domes, volcanic cones, and basaltic sheets. In Cyrenaica and Marmarica, the elevations of Jebel Al Achdar and Jebel Al Akaba are the northern flank of an asymmetric flat uplift of the limestone layers, which change to the south into the wide Qattara depression (−133 m) (Fig. 4.10).

The limestone and sandstone cuestas and scarped tablelands (mesas) of the Libyan Deserts extend via the Nile Valley to the

**Fig. 4.8** The cuesta landscape in the framing of the Murzuk Basin, Central Sahara. *Photo* NASA, Gemini Mission 5 (1965)

**Fig. 4.9** Eastern side of the Aïr Mountains at Adrar Chiriet to the Ténéré desert, Rep. of Niger. *Photo* R. Baumhauer

**Fig. 4.10** The cuesta of Bilma in the southern central Sahara, developed in Late Cretaceous sedimentary rocks, Northeast Rep. of Niger. *Photo* R. Baumhauer

east. The Atbai, a fault block mountain made of crystalline schists, old eruptive and intrusive masses, is heavily eroded, and the waters coming from it dissect the plateau of the western foothills into a gentle highland. It merges south into the peneplains of the Eritrean highlands.

West Africa and thus the **Sahel**, **Sudan zone**, and **Guinea coast** are part of **Lower Africa**. They are divided into three large geological basins, separated from each other by ridges. From east to west, these are the Upper (or White) Nile and the Chad Basin, separated by the Darfur Ridge, and the Niger Basin as parts of the Precambrian basement complex of Africa. The characteristically wide and several hundred meters elevated ridge areas are the first-order morphological characteristics of Lower Africa. However, because the ridges rise very gently over the basins, the morphological structure of Lower Africa is less clear than that of the rest of Africa. To the south, the basins are bounded by the Upper Guinea Ridge, which only reaches heights up to 1,500 m in individual elevations (see Fig. 4.4). The Precambrian basement consists predominantly of granites, gneisses, and schists, which form inselbergs around the ridges. In the Paleozoic sandstones covering the basement, typical cuesta and mesa landscapes have formed. Active sand and dune areas are widespread in the northern parts of the Sahel, such as the sand sea of El-Djouf in Mauritania or the Great Erg of Bilma in the Rep. of Niger. In the southern Sahel, the (old) dunes today are fixed by vegetation. They were active in the last ice age about 18,000–20,000 years ago, for example, north of Bamako in Mali and Ouagadougou in Burkina Faso or southern Niger from Maradi via Zinder to Lake Chad and are still preferred grazing areas for the cattle herds of the nomadic Fulbe (Fula, French: Peul). In years with more rainfall, millet and sorghum are also grown on the old dunes in the southern Sahel, which, especially in recent decades, due to the increasing pressure of use as a result of the rapid population growth, has repeatedly led to partly violent conflicts between the sedentary farmers and the nomadic cattle herders (see Chap. 7).

The lowlands of Senegal and Gambia to the middle Niger River are developed on flat-lying or gently tilted Paleozoic sandstones, shales, and limestones with an elevation below 200 m. Only in the extreme southeast, in the Niokolo-Koba National Park and Bassariland, the foothills of the Fouta Djallon rise as a hard rock massif of the African Precambrian basement to almost 500 m. The headwaters of the Niger River and the Senegal River flow through this in narrow valleys. While the Senegal River turns west and deposits a vast alluvial plain on the coast, the Niger River flows in an arc, the so-called Niger Bend, with an ever-decreasing longitudinal gradient to the northeast into the Sahel zone and the southern Sahara. There the river branches off between the cities of Ségou and Timbuktu (Tombouctou) in Mali into several arms and, together with its tributary Bani, develops the Massina (also Macina), a floodplain of around 40,000 $km^2$, the Niger inland delta, in which the Malian cities Djenné on the Bani and Mopti are located. The present Massina has been part of a roughly 300,000 $km^2$ large endorheic basin since the Late Pleistocene, which was last occupied by a closed lake 11,000 years ago. Since the Middle Holocene 5,000–8,000 years ago caused by a river capture at Tobaye (Mali) from the southeast by the present Niger lower course, the Niger River drained into the Gulf of Guinea. Large parts of Burkina Faso and the western and southern parts of the Rep. of Niger are dominated by wide plains at 200–300 m. These peneplains of the Precambrian basement are only interrupted by inselbergs and low Petroplinthite (laterite) plateaus. Scattered cuestas and scarped tablelands of Paleozoic sandstones are superadded. The best-known examples are the Manding plateau in southwestern Mali or the impressive Cuesta (Falaise in French) of Bandiagara of the Gandamia Mountains rising to 1,080 m, with their inselbergs near Hombori east of Mopti in Mali or the cuesta of Banfora, which continues to Burkina Faso. Bandiagara is also known for the impressive buildings of the Dogon refuge settlements (Fig. 4.11).

The basins of Dallol Bosso in western Niger and the Chad Basin are separated by a southern

**Fig. 4.11** Falaise (cuesta) de Bandiagara in the Mopti region of Mali. *Photo* Torsius, Adobe Stock

continuation of the Aïr Ridge. The Dallol Bosso Basin is developed on flat-lying Cretaceous and Eocene limestones and sandstones. To the south, the Jos Plateau follows, a gneiss plateau dominated by granitic inselbergs. Morphologically also indistinctly separated from the Niger Basin is the **Chad Basin**. About 100 m thick Cenozoic sediments give the Chad Basin a monotonous landscape, which shows a widespread alluvial plain, partly covered by fixed dunes. **Lake Chad**, with a mean depth varying between three and five meters, currently contains water in its southernmost part only. The free water surface is less than 2,000 km$^2$. The lake, which is not located at the deepest point of the Lake Chad Basin, is fed by the Schari and Logone rivers, the only perennial rivers of Chad. They have their sources in the plateaus of the Central African Ridge, which together with the foothills of the Adamawa Highlands, bound the Chad Basin to the south. Both rivers merge about 100 km south of Lake Chad and thus slightly south of the Chadian capital N'Djamena. Before the sedimentation of the dry bed of the Bahr al-Ghazal River, the two rivers flowed into the deepest part of the Chad Basin, into the Bodelé depression. In very wet periods, such as in the years 1854, 1866, and 1874, the waters of Lake Chad penetrated up to 150 km into the Bahr al-Ghazal. On the other hand, multi-year drought periods, including in the eighties of the last century, 1973, 1914 or 1904 led to an extreme shrinkage to a few square kilometers of the large residual lake, which only existed in the immediate mouth area of the Chari. The northern border of the Chad Basin is occupied by the extensive Saharan sandstone plateau of the Ennedi and the volcanic Tibesti Mountains. In the east, the Chad Basin is bounded by the plateau of Ouddai (Wadei) as part of the Darfur Ridge. It extends in the southeast with the massifs of Abou-Telfane (1,360 m) and Guéra (1,800 m) into the Chad Basin.

From the eastern border of the Chad Basin to the western slope of the Highlands of Ethiopia, the relief is determined by the extensive peneplain of the Sudd. This wetland area in South Sudan, formed by the White Nile, is one of the largest swamp and floodplain areas worldwide. In the Darfur Ridge occurs, the young volcanic massifs of Jebel Marra (3,024 m), hogbacks, and (granitic) inselbergs. To the north, the dissected sandstone plateaus of Ennedi and Erdi lead into the Central Saharan Ridge. In Kordofan, east of Darfur, the prominent cuestas developed in the Nubian Sandstone of the (White) Nile Basin begin.

The Upper Guinea Ridge, a low uplift, divided into several sections by transverse depressions, shapes the relief of the Sudan zone and Guinea coast. Although it only has elevations up to 1,500 m, extensive plateaus have formed in the overlying Paleozoic rocks, such as the Jos Plateau in Central Nigeria (1,400 m). In the Togo-Atakora Mountains, the Precambrian basement has been exposed, to which a scarped tableland of Paleozoic sandstones follows to the west in the Oti-Volta area. Greater altitudes are only reached by the Liberian Shale Mountains in Nimba (1,850 m) and the Guinean highlands of Fouta Djallon (1,515 m), which are developed on Paleozoic sandstones with strong diabase intercalations. Mountainous regions and peneplains extend in some sections to the coast. The typical morphological complex on the basement ridge is characterized by a gently undulating peneplain-inselberg relief. At the transition to the next lower peneplain, distinct escarpments occur, often several hundred meters above the ground. In addition to the Jos Plateau, such peneplains with pronounced cuestas occur in the Idanre Mountains in southwest Nigeria, on Mount Agou in Togo, and in the Nimba Mountains in the border region between Guinea, Liberia, and Ivory Coast (Republic of Côte d'Ivoire).

The massifs of Dan and Toura, as offshoots of the Fouta Djallon, reach heights up to 1,300 m (Mt. Momi). The highlands of Fouta Djallon from Guinea via Sierra Leone to Liberia limit the 30–80 km wide coastal zone to the east. South of the Senegal delta, the dominant steep coasts consist of Precambrian and Paleozoic sediments. The cliff coast alternates with a few mangroves, and flat coastal sections.

In Guinea, the Fouta Djallon highlands are a barrier between the coastal zone and Upper Guinea, which climatologically and geomorphologically already passes over to the Sudan zone. The Fouta Djallon reaches heights up to 1,515 m and is divided by deeply incised gorges into a series of plateaus. The Niger River and its tributaries have cut into the approximately 400 m high plateaus of Upper Guinea that follow the Fouta Djallon to the east. Hardened to crusts (Petroplinthite) Plinthosols, and Saprolites are characteristic of these treeless highlands.

Between the coastal plain and the Guinea Highlands lie in Sierra Leone and Liberia the so-called inner plains. They consist of a peneplain at around 120 m a.s.l. with a few superimposed hill chains of an older erosion surface: Malal, Kasewe, Moyamba, and Mokanij Hills. Bauxite has been mined in the Mokanij Hills of the inner plains since the mid-sixties. In Marampa, Sierra Leone, and the Liberian part of the Nimba Hills directly on the border with Guinea and Ivory Coast, iron ore is in the form of hematite (see Chap. 3).

In the Republic of Côte d'Ivoire and Ghana, a flat peneplain, which extends over 700 km into the Sudan zone to the north, shapes the relief. In Ivory Coast, the peneplain reaches a height of 200 m in the northern parts of the country from 350 to 500 m are reached. These peneplains are only exceeded by isolated granite inselbergs. More relief intensity can only be found in the extreme northwest of the country. In Ghana, whose 500 km long coast is referred to as the "**Gold Coast**" (and in the time of colonialism gave the country its name, see also below). About 100 km east of the capital Accra, the large cuspate delta of the Volta River seems to be a direct outgrowth of a natural change in the location of the mouth of the Volta and of a marked reduction in sand supply. The Volta River is, next to the Senegal River and Niger River, the third major West African river. 100 km north of its mouth since 1966, the Akosombo Dam has been damming the Volta River in a gorge of the Akwapin Mountains (named in Benin as Atakora Mountains). In the Volta Reservoir, which extends over 520 km to the city of Yapei in northern Ghana, the White Volta and the Black Volta, the Afram, Daka, Pru, and Oti flow directly into the lake. The Red Volta, the Nasia, and the Kulpawn flow into the White Volta before they enter the lake. The lake's area is about 8,500 $km^2$ (for comparison, Lake Constance is 538 $km^2$), and its mean depth is between 18 and 19 m. The reservoir volume has a capacity of around 153 billion cubic meters, which corresponds to 153 $km^3$. Thus, in terms of reservoir volume, Lake Volta is the fifth largest, and in terms of its area, the second largest reservoir on earth.

Geologically and morphologically more varied than the Ivory Coast (Republic of Côte d'Ivoire) and Ghana are the two states to the east of Ghana, Togo and Benin. Togo has a width of only 50 km at the Bay of Benin, and Benin has a width of 125 km. Togo extends to the border of Burkina Faso about 600 km to the north, Benin even extends to its northern border with the Republic of Niger over 760 km. The barrier spit coast of both states comprises a complex system of lagoons and coastal lakes (for example, Lake Togo or Lac Nokoué). Inland from the seaside are the plateaus of the Terre de barre, an intensively farmed plain with sandy-clayey soils rich in organic matter (Lixisols, Nitisols, Vertisols, see Chap. 7). They extend about 70 km north in Togo, and even about 130 km in the east of Benin, reaching as far as the southwest of Nigeria south of Abeokuta. To the north of the Terre de Barre is the Upper Guinea Ridge, which is called in the Togo Plateau of Central Togo and the Benin Plateau of Central Benin. It extends to about 10° N, where it is bounded to the northwest by the foothills of the Atakora Massif and to the northeast merges into the Niger plain of North Benin. The plateaus, which are already part of the savanna woodland, are peneplains, which are exceeded by 250 m tall inselbergs (Bornhardt), among others at Dassa-Zoumé. The Plateau of Central Benin is bounded to the northwest by the Atakora Massif, which is part of a mountain range made up of Precambrian rocks, which extends from the southeast of Ghana over the well-watered and still small-scale farming-dominated Togo mountains. With mean elevations of 600–700 m, it is Togo's most important cocoa and coffee growing area and continues the Kabre mountain country in Central Togo to the northwest of Benin. In Togo, the mountainous region of North Togo extends to the Oti lowlands and the Dapaong plain at the transition to the thorn tree and tall grass savanna.

Along the simplification coast on the Gulf of Guinea, there are beach ridges and a multitude of lagoons. Most seaports on the West African coast, such as Monrovia in Liberia, Lomé in Togo, or Cotonou in Benin, are artificial ports. The **Niger Delta**, one of the main oil-producing areas in Africa, covers 28,827 km$^2$, of which 19,135 km$^2$ are above sea level. The delta plain is characterized by floodplain areas, mangroves, and barrier beaches. South of Mt. Cameroon on the west coast of Africa is characterized by lagoons, marsh areas, and barrier beaches and there are many narrow basins separated by basement rocks and Cenozoic sediments.

Morphologically very monotonous is the central (or middle) African **Congo Basin** and its peripheral uplift areas, which are part of the characteristic ridge structure of Africa and formed during the rifting processes when Gondwana broke apart in the Paleozoic. Most recently, the uplift tendencies of the ridges intensified in the Oligocene (Ch. 3). The subsequent erosion processes in the Miocene and Pliocene created the two Post-African Surfaces, which are particularly well observed in the ridge areas. Where granite is exposed, either a typical peneplain relief occurred by the uplift and subsequent erosion, or the relief has been so modified by fracture tectonics that the typical tectonic graben and horst structures have been developed. A characteristic graben with pronounced volcanism has been formed since the late Cretaceous by the **Cameroon Line**. Starting from the Mandara Mountains in northern Cameroon, this tectonic zone continues southwest via the Alantika Mountains, the Adamawa Highlands, the Bambouto and Manengouba Mountains to Mount Cameroon (4,070 m), an active volcano, and further over the islands of Fernando Poo, Príncipe and Sao Tomé to Annobónu in the Gulf of Guinea. The Adamawa Highlands consist mainly of undulating peneplains, which are dominated by inselbergs of granite. The Cameroon Highlands drop off quite steeply to the Benue fault trough, which separates the highlands from the Upper Guinea Ridge. As a result of the continental tectonic fragmentation processes that have been ongoing since the Mesozoic, not only uplift and shearing movements, but also volcanism occur in the area of the Central African Ridge and the East African Rift. Examples of this are the heaved block (horst) of the Ruwenzori Massif (5,110 m) and the Virunga Volcanoes (4,507 m). In addition to these volcanic and tectonic relief forms on the eastern peripheral ridge of the Congo Basin, well-preserved Miocene, in parts possibly even older peneplains and escarpments, as well as bevelled fault blocks dominate the landscape of the ridges of the northern, western and southern borders of the Congo Basin, which consist of sandstones and Precambrian shales. Asande- and Lower Guinea Ridges meet in the Cameroon Highlands. As a result of strong tectonic activity, it is a mosaic of different downthrown blocks of metamorphic shales, old, predominantly granitic intrusive bodies, and young volcanic. In the Lower Guinea Ridge, the denudation of old fold structures of the basement led to the formation of morphologically only slightly pronounced mountain ranges. The southern part of the ridge, which extends into Angola, has a sequence of peneplains and escarpments that is largely like the Great Escarpment in southern Africa. In the basin itself, very flat Mesozoic sandstone rocks sloping towards the interior have created an extensive cuesta landscape, and in the deepest part of the basin, Pliocene to recent deposits have created wide plains. The Central African Coastline (Lower Guinea), from the Biafra coast of Nigeria to the mouth of the Congo into the Atlantic, contrasts with the Guinea coast of West Africa due to the coast-parallel Lower Guinea Ridge on the one hand, very narrow. On the other hand, the Lower Guinea coast has a more diverse configuration of the coast. In addition to a simplification coast with spits, lagoons, and the delta of the Ogooué River in Gabon, wide estuaries are also formed. The coastal section with the Congo River mouth is part of a ria coast. The Congo drains in a submarine canyon up to 2,700 m below sea level. South of its mouth, steep coast areas with up to 50–150 m high cliffs prevail. Cliffs can also be found in Cameroon and Congo. The coastal foreland itself is only slightly relieved, reaches a width of 30–200 km, and heights up to about 300 m, and consists of marine and continental sediments from Jurassic, Cretaceous, and Cenozoic. The rocks consist of sandstone and limestone with oil and natural gas deposits as well as potash salt, asphalt, and lead deposits. Mangroves can be found throughout the coastal area.

## 4.3 The Geomorphology of High Africa

Within Africa, the first-order morphological characteristic is the broadly elevated east and south of the continent. In contrast to Low Africa with a mean elevation of 300 m a.s.l., High Africa has a mean elevation of 1,200 m a.s.l. (see Fig. 4.1). It includes the Highland Massif of Ethiopia, the Highlands of East Africa, the Zambezi lands south of the Congo Basin, the mountain massif of Lesotho with heights up to 3,400 m, the Kalahari (high) Basin, which at its lowest point in Botswana has a sea level of 1,000 m, the dry areas on the South Atlantic and the South African Phanerozoic Cape Fold Belt. As in Lower Africa, the geological ridges and basins also determine the relief in High Africa.

Due to its high altitude and the distinctive **Rift Valley**, the East African part of High Africa exhibits a more pronounced and finely differentiated physical-geographical structure. The Cenozoic rift system in Eastern Africa encompasses the **Southern Rift, the Central African** and the **East African Rift**, and extends from southern Mozambique north to Djibouti and then continues into the Red Sea. Associated volcanic activity, initiated at the end of the Jurassic and intensifying since the Oligocene, resulted from intraplate volcanism driven by the activation of the Afar-, Karoo-, and Walvis-(Super mantle-)Plumes. In East Africa, particularly the Afar-Plume, led to strong and large-scale uplift and deformation processes throughout Africa, which in East Africa led to fracturing in the existing rock along existing zones of weakness and thus also to the formation of the East African rift system (see Chap. 3). It can be divided into four zones: Close to the Equator it is made up of eastern and western rifts to either side of the Victoria Basin. The **Western Rift** extends from Lake Albert (from 1972 to 1997, also known as "Mobutu-Sese-Seko-Lake") to Lake Tanganyika; the **Central African Rift** has younger forms, rugged scarps and younger dissections than the East African Rift; the Eastern Rift in Kenya and Tanzania, including the branches where Lake Eyasi and Kavirondo Bay are located. The **Ethiopian Rift Valley** follows to the northeastern: from the Afar Depression to Lake Turkana, and to the south, the **Malawi Rift Valley**, which extends from Lake Malawi (Lake Nyasa) to the Shire Valley. The rift system extends from the Afar Depression in the north to beyond Lake Malawi in the south, about 5,600 km.

Apart from **Lake Victoria**, the large lakes of East Africa lie in the rift valleys. They are linearly shaped, very deep, and bounded by fault scarps. The bottom of Lake Malawi is up to 700 m below sea level. On both sides of the rift valleys and the valley floors themselves, there are hundreds of volcanoes, most of which, however, are quiescent volcanoes. In some parts of the Red Sea Rift, in the Aden Rift, and the northern and eastern Afar Depression, ocean floor spreading is currently taking place.

The **Eastern Rift Valley (Eastern Branch)** runs from the Afar Depression on the Red Sea coast through central Ethiopia and Kenya with characteristic volcanic complexes. The northernmost section of the Eastern Rift Valley is the Afar Depression, a roughly triangular lowland with a north–south extent of more than 600 km, from the Gulf of Zula on the Red Sea (Eritrea) to the Ethiopian Rift Zone (9° N), and a west–east extension of about 400 km, from inland Djibouti to the Gulf of Aden. In the Afar Depression, the valley floor is currently widening by a few centimeters per year, accompanied by seismic and volcanic activity. The depression is bounded by fault scarps, which rise to 3,000 m to the contour line of the Ethiopian plateau in the west and 2,000 m to the plateau in the south. Further south, the Ethiopian Rift Valley connects. It is 60–70 km wide and stretches in the north in a southwestern direction and the south south-southwest and is bounded by fault scarps up to 1,000 m thick, such as the Guraghe Escarpment in Ethiopia. At other places, the rift shoulders/rift boundaries are less clearly defined. East of Addis Ababa, the capital of Ethiopia, the rift shoulder presents a wide, gently downwardly curved form. East and west of the fault zone, basalt plateaus up to 4,000 m thick are formed and both on the rift floor and the rift shoulders, shield volcanoes of different ages are present, including trachytic volcanoes east of the rift valley (Kaka, Chilalo, Hunkuolo) (Fig. 4.12).

To the south of the Afar Depression, the approximately 1,000 km long **Gregory Rift Valley** connects. It runs in a north–south direction, is 60–70 km wide, and extends from Lake Turkana to northern Tanzania (Lake Natron Basin). To the west of it is the Nyanca Rift Zone, which initially runs east–west and then northeast-southwest and after 130 km transitions into Lake Victoria. Although some stratovolcanoes are found on the rift floor, however, the former volcanic activity and its products are not comparable to that of Ethiopia. Some of the volcanoes are located 50 km or more from the current rift boundary. Among them are the direct basement sitting volcanic massifs of Kilimanjaro (5,895 m), which was last active 36,000 years ago, Mount Kenya (5,194 m), and Mount Elgon (4,322 m), which are the highest elevations. Volcanic lava plains and plateaus have partly been transformed into deeply dissected mountainous countries at the edges of the trench. At 2° S, the tectonic structure and the magma type change. The well-defined trench formation is weakened and is now referred to as Tanzania Divergence. On the Tanzanian Craton, the rift zone disappears but continues further southeast in the **Pangani Trench** (Fig. 4.13).

The **Western Rift (Western Branch)** begins at Lake Victoria and runs through Lake

**Fig. 4.12** Eastern rift valley (Eastern Branch) near Narok (Narok County) Kenya. *Photo* R. Patzel, Mauritius

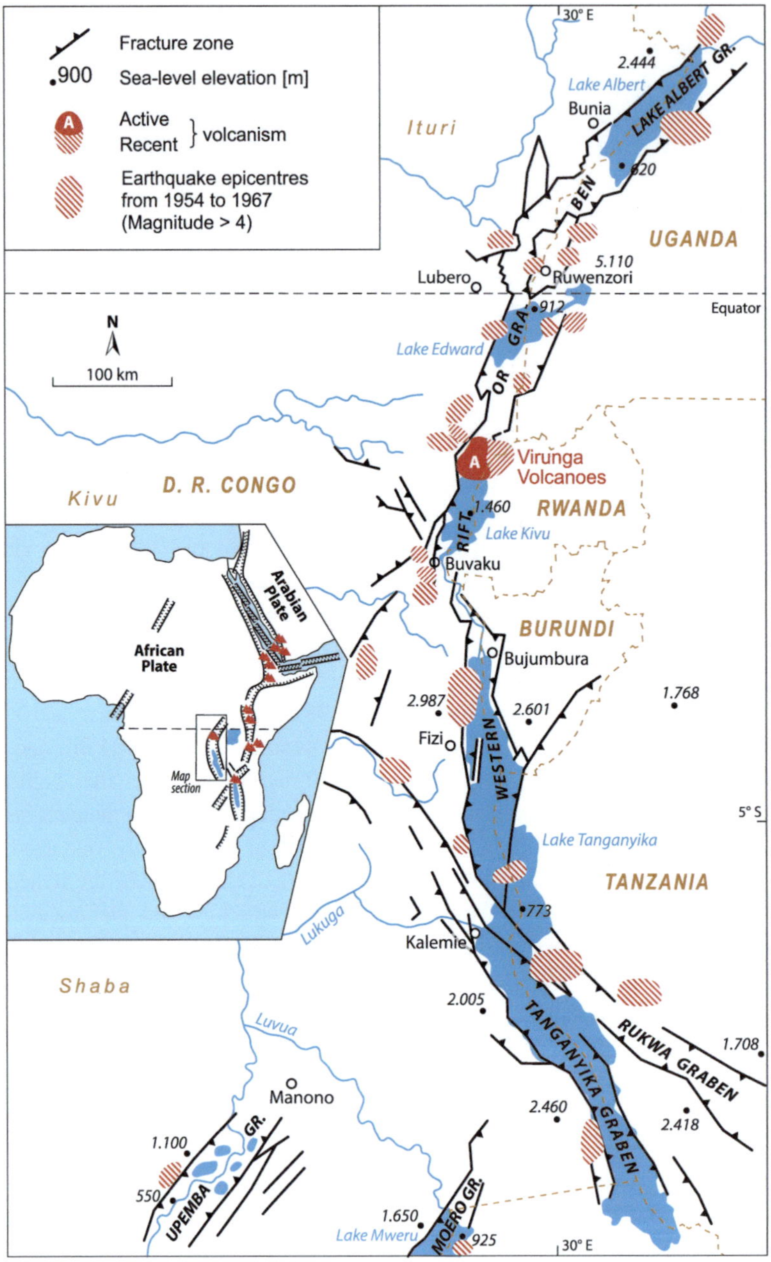


**Fig. 4.13** Western rift valley (Western Branch) of the East African Rift Zone. Modified after Wiese (1997)

Tanganyika and Lake Malawi to the southern coast of Mozambique. Four sections or volcanic provinces can be distinguished: Toro-Ankole, Virunga, South Kivu (north of Lake Tanganyika), and Rungwe (at the northern end of Lake Malawi). The northernmost area is the **Albert Rift Zone**, which starts north of Lake Albert and extends to the South Kivu volcanic province. Northeast-oriented faults define the course of the White Nile. At about 4° N, the Albert Rift Zone is interrupted by a Precambrian basement structure oriented to the northwest, the **Aswa Shear Zone**. South of Lake Albert is the Semliki Basin in the west, and between Lake

Albert and Lake Edward (sometimes referred to as "Lake Amin" or Lake Rutanzige), the Ruwenzori Mountains between the DR Congo and Uganda. It consists of a block of a raised Precambrian basement, and a horst of over 5,000 m height is Africa's highest non-volcanic mountain range. It is divided into six main peaks, which are glaciated at altitudes above 4,200 m. Non-volcanic mountains can also be found east of the western rift: the Eastern Arc Mountains in East Tanzania, whose precursors were probably formed before 100 Ma. The current mountains, however, are due to tectonic activity, the general elevation of the Central African Ridge in the Middle of Palaeogene before 35–25 Ma (Oligocene).

Further south in the border area of the DR Congo, Rwanda, and Uganda, the western rift crosses the basin of Lake Edward and the Virunga Volcanic Province. It consists of basaltic shield volcanoes of Miocene to recent age. Following the Virunga Volcanic Province is the Lake Tanganyika Rift Zone. Here are only half-graben structures, at least ten basins, but no magmatism/volcanism present. In the basin of Lake Tanganyika, a vertical displacement of geological layers with fault throws of up to 6,000 m has been detected. The current height differences between these displacements are significantly lower. They are due to the erosion of the valley flanks and the subsequent sedimentation of the material on the valley floors, as well as the accumulation of up to 1,800 m thick volcanic lavas on the original valley floor.

Parallel to the **Tanganyika Rift Zone** is the **Rukwa Rift Zone** in Tanzania at 50–100 km. It is about 350 km long and 50 km wide and is bounded in the west by the Ufipa escarpment with an elevation of 1,000 m and in the east by the flatter Lupa escarpment. To the southeast, the Rukwa Rift Zone crosses the Rungwe Volcanic Province, which is located north of Lake Malawi. The Lake Malawi Rift Zone is 800 km long and has a width between 40 and 90 km. It shows half-graben structures like those at Lake Tanganyika and is divided into nine basins. To the south, over 600 km, the **Urema Graben** and the **Dombe Trough** follow.

Due to the multiple changes of periods of relative uplift calm and increased uplift occurring in the Cenozoic, around the entire **South African highlands**, from the Limpopo to the Kunene, the border river between Angola and Namibia, the formation of escarpment has occurred. The Great Escarpment of South Africa is the upper uplifted block because of a particularly intense initial uplift in the Oligocene. The highest elevation, the Basuto highlands of the Drakensberg, is a hard rock massif of old volcanic masses. The dissection of the old, layered plateaus results in the formation of mountainous regions with the characteristic forms of table mountains, as in the Waterberg, in Namaqualand and the Kaokoveld. The Precambrian basement complex can be found in the highlands of Transvaal (since 1994/2002 the South African provinces of Limpopo Gauteng, Mpumalanga, and North-West), in Griqualand West, and the Windhoek highlands. Large parts of the plains of Upper Karoo, the Oranje, Bushman, and Namaqualand to the Kalahari Basin get their morphological character from the Karoo sediments and form a cuesta landscape with table mountains, monadnocks, and dolerite relict mountains. Aside from **peneplains and pediplains cuesta landscapes** and **structural plateaus** represent another type of relief that is very widespread in Africa. Cuestas are built from bottomed positioned sedimentary rocks of different hardness. In some regions, the denudation of the Karoo sediments shows well-preserved traces of the Late Carboniferous-Early Permian glaciation, e.g., striations, glacial polish, grooves on bedrock (Figs. 4.14 and 4.15).

Young sediments of great thickness fill the Kalahari Basin, whose landscape is characterized by large dune fields and windblown sand fields. Under the current climatic conditions, the dunes are largely fixed by vegetation. From the western ridge, dry valleys lead into endorheic basins, where the periodically incoming water evaporates. Among them, the Makgadikgadi pans and the **Etosha Pan** in Botswana and Namibia are the largest. The Lunda Ridge as the northern boundary of the Kalahari Basin drops to the south in a landscape of structural plateaus dissected by

**Fig. 4.14** A Hundred-meter-tall inselberg in a peneplain of the same rock, Spitzkoppe, Central Namibia. *Photo* J. Kempf

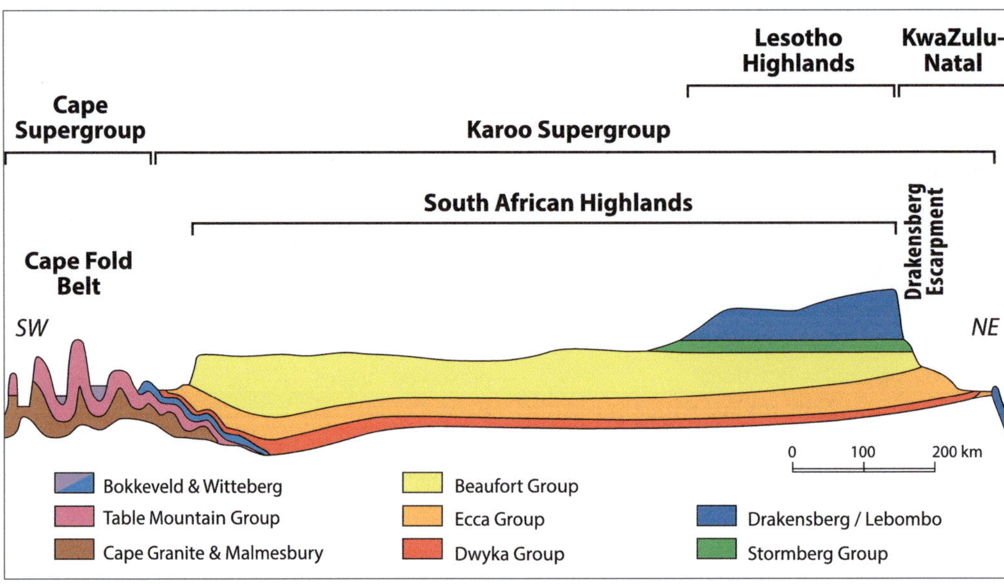


**Fig. 4.15** Geological SW-NE profile across the Republic of South Africa. The Table Mountain Group is part of the Paleozoic Cape Supergroup; Dwyka-, Ecca-, Beaufort-, Stormberg-Group form the Karoo Supergroup. Modified from Catuneau et al. (2005)

numerous valleys. Here, many perennial (water-bearing all year round) rivers rise, of which only the Okavango River drains endorheically into the basin named after it. The other rivers drain to the Zambezi River, which crosses the eastern ridge in a depression.

The Limpopo and Oranje Rivers, which rise in the eastern highlands, behave differently: While the Limpopo turns east, the Orange River flows over the Cape Plateau to the west and transverses the flat western edge ridge in the South African Northern Cape Province near Upington (Fig. 4.16).

The **Namib Desert**, extending over 2,000 km along the west coast of southern Africa, is bounded to the east by the Great Escarpment. The Namib is about 120–200 km from the coastline and represents the western boundary of the African plateau. Large parts of the Namib consist of rocky or sand-covered plains with some flat hill ranges and isolated inselbergs. The sand sea of the Namib covers about 34,000 km$^2$, slightly more than 41 per cent of the area. The Karoo, a semi-desert landscape of rocky plains and low hills of Mesozoic sandstone, shale, and tillite, which covers almost a third of the area of the Republic of South Africa, extends from the Cape Fold Belt in the south, north to the Orange River at an altitude between 900 and 1,200 m. In the western and northern peripheral areas, the Precambrian basement complex is exposed at the surface.

The 2,100 km long simplification coast on the Indian Ocean is quite flat, with long sandspits covered by dunes, which are in front of shallow lagoons, marshes, and swamps. In central Mozambique, the coastal plain is up to 300 km wide, and only north of the Mocambo Bay at 15° S, a cliff and coral-reef coast appear more frequently. Along the Somali Peninsula, aeolian deposits increase. Aeolian sands, lagoon evaporites, and coral reefs also characterize the coastal areas on the Red Sea.

## 4.4 The Cape Province of South Africa

The mountains of the western Cape are in front of the Great Escarpment and consist of several wide valleys interrupted by successive fold mountains (Cedarberg ranges, Swartberg) and

**Fig. 4.16** Overgrown old red sandy longitudinal dunes in the Kalahari, Southeast Namibia. *Photo* D. Busche

the intervening Tankwa Karoo. The highest areas achieve heights up to 2,078 m, with the Piketberg in the hinterland of Cape Town. The **Cape Fold Belt** is significantly older, but also smaller than the Mediterranean alpine Atlas ranges in northwestern Africa. The mountain building began in the then southwestern border area of Gondwana through the collision of the (Proto-)African plate and the South American plate 280–230 Ma at the turn from the Paleozoic to the Mesozoic. The folding took place in the Triassic and the Cretaceous, the uplift with the faulting tectonics in the Tertiary, and the dissection in the Neogene (Quaternary). Today's landscape is characterized by high mountain-like crests and summits in Table Mountain sandstones stratigraphically above a granitic basement. Earthquakes also occur along the fault lines.

The Cape Coast, a cliff coast between the Lepelle and Tugela Rivers, consists of Precambrian and Paleozoic rocks as well as thick deposits of the Karoo Formation. The Karoo Supergroup forms powerful cliffs further to the northeast (Fig. 4.17).

## 4.5 The Current Morphodynamics

Looking at the present landform development, Africa shows a sculptural, climatic geomorphological zoning of relief types in a meridional direction. It superimposes the structural division into ridges and basins.

All regions of Africa share the existence of ancient peneplains, formed under a wet savanna climate with periodic water flow (see Chap. 5). The landscape of the wet-dry tropics is characterized by the interaction of intense chemical weathering and dominant fluvial shaping with periodic runoff, which in turn also has a high degree of areal degradation. The predominant valley form here is the

**Fig. 4.17** The Outeniqua mountains are part of the South African Cape Fold Belt (Western Cape Province, South Africa). *Photo* Helge, Adobe Stock

flat-bottomed valley. The mountains rise with steep slopes from the plain. This landscape is characterized by the wide distribution of **peneplains and inselbergs**. Equatorward, in the transition to the wet-equatorial tropics, the more abundant and longer-lasting rainfall causes a perennial runoff of the waters, which leads to a stronger incision of the valleys. The increase in fluvial activity occurs in the formation of V-shaped valleys. Mass movements such as landslides and slow creeping of wet soil because of strong soil saturation are significantly involved in the current morphological processes.

If we summarize the currently dominant relief types of Africa, then it is the **peneplain-inselberg relief** with the associated pediments and table mountains sealed by Petroplinthite (laterite) that are the defining landscape type of tropical Africa. However, the processes responsible for their formation are by no means uniform. In arid and semi-arid areas, **pediments** are formed at the mountain rims by the effects of mechanical weathering and episodic water flow, which gradually grow together to larger plains. In the wet-dry tropics, chemical weathering produces a mantle rock, into which the debris material eroded from the steeper slopes is also incorporated. Although the peneplains of Africa are predominantly Cretaceous to Cenozoic landscapes of past ages, they are also continuing to develop, depending on the predominant climatic conditions. Table mountains with Petroplinthite crusts (laterite) are witnesses of dissolved peneplains and currently only exist as relics in East Africa and Sudan.

In addition to the peneplain landscape, the fault scarp and fault block landscape, and the volcanism in the rift systems, especially in East Africa, **cuestas, scarped tablelands,** and **structural plateaus** are particularly widespread in the tectonic depressed and uplift areas of West Africa, and the Sahara. In the semi-arid and arid areas of the subtropics, in the Sahara, the Kalahari, and the Namib (which, as a coastal desert, is a special case), there is strong mechanical weathering. Episodic rain events produce heavy runoff, giving rise to channel incisions and denude very extensive before the water collects in the dry valleys (Wadis, Ennedis, Rivieren). Slope wash and rock falls are added at the escarpments, scarps, and cuestas. In the Kalahari, the dunes are currently fixed by vegetation, which significantly reduces the effectiveness of the aeolian denudation.

## References

Ahnert F (2015) Einführung in die Geomorphologie, 5th edn. Ulmer, Stuttgart, p 393

Bremer H, Sander H (2000) Inselbergs: geomorphology and geoecology. In: Porembski S, Barthlott W (eds) Inselbergs: biotic diversity of isolated rock outcrops in tropical and temperate regions. Ecological studies, vol 146. Springer, Berlin Heidelberg

Catuneau H, Wopfner PG, Erikson BS, Rubidge R, Smith MH, Hancox PJ (2005) The Karoo basins of south-central Africa. J Afr Earth Sci 43:211–253

Davis WM (1899) The geographical cycle. Geographical Journal 14:481–504.

King LC (1942) South African Scenery. A textbook of geomorphology, Oliver and Boyd Edinburgh, p 308

Manshard W (1981) Afrika—südlich der Sahara. Fischer, Frankfurt, p 327

Wiese B (1997) Afrika: Ressourcen, Wirtschaft, Entwicklung. Teubner, Stuttgart, p 262

## Further Readings

Baumhauer R (2024) Physische Geographie—Geomorphologie, 4th edn. Borntraeger, Stuttgart, p 180

Besler H (1992) Geomorphologie der ariden Gebiete. – Erträge der Forschung Bd. 280. WBG, Darmstadt, p 189

Billi P (ed) (2022) Landscapes and landforms of the horn of Africa. Eritrea, Djibouti, Somalia. Springer, Berlin, Heidelberg, p 403

Boardman J, Holmes PJ (eds) (2018) Southern African landscapes and environmental change. Taylor & Francis, Milton Park, p 328

Bremer H (2004) Water movement, landform and soil structures in the humid tropics. Ztschr Geomorph 136:135–172

Bridges EM (1990) World geomorphology: Africa. Cambridge University Press, Cambridge, pp 30–65

Buckle C (1978) Landforms in Africa. An introduction to geomorphology. Longman, London

Burke K, Gunnell Y (2008) Th African erosion surface: a continental-scale synthesis of geomorphology, tectonics, and environmental change over the past 180 million years. The Geological Society of America, Memoir, p 201

Guillocheaua F, Brendan S, Babya G, Bessina P, Robina C, Dauteuila C (2018) Planation surfaces as a record of mantle dynamics: the case example of Africa. Gondwana Res 1819:62
Heine K (2019) Das Quartär in den Tropen. Springer, Berlin, p 754
Hipondoka M, Mauz B, Kempf J (2013) Chronology of sand ridges and the Late Quaternary evolution of Etosha pan. Geomorphology 204:553–563
Kempf J (2010) Morphotectonics and denudation in the landform history of the Namibian Great Escarpment landscapes. Ztschr Geomorph 54:347–376
Lancaster N (1996) Desert environments. In: Adams WM, Goudie AS, Orme AR (eds) The physical geography of Africa. Oxford University Press, New York, pp 211–237
Meadows M, Holmes P (eds) (2012) Southern African geomorphology. African Sun MeDIA, Stellenbosch, p 429
Moore A, Blekinsop T, Cotterill F (2009) Southern African topography and erosion history: plumes or tectonics? Terra Nova 21:310–315
Partridge TC (2010) Tectonics and geomorphology of Africa during the Phanerozoic. In: Werdelin L, Sanders WJ (eds) Cenozoic mammals of Africa. University California Press, Oakland, p 998
Runge J (ed) (2010) African palaeoenvironments and geomorphic landscape evolution. Palaeoecology of Africa, vol 30. CRC Press, Baton Rouge, p 336
Skowronek A (2020) Wie entstehen Abtragungsflächen? Reliefverebnung durch Pedimentation. Mitt Österr Geogr Gesell 162:211–254
Thomas MF (1994) Geomorphology in the tropics. A study of weathering and denudation in low latitudes. Wiley, Chichester, p 482

# 5 Climate: A First-Order Factor in the Physical Geography of Africa

Due to its nearly symmetrical location with respect to the latitude, Africa's northern (37° N Cape Blanc) and southern (35° S Cape Agulhas) extremities extend almost equidistant from the Equator. Broadly similar series of climates can be traced northward and southward from the hot and moist Equator belt. The Sahara has its counterpart in the arid tracts of the Namib and Kalahari, Sudan in Zimbabwe, and the Mediterranean coast in the southwest of South Africa. Africa's climate is largely subject to tropical circulation processes. Only the most poleward regions of the continent, North Africa and the Cape, are influenced by the mid-latitude westerlies and their associated disturbances. Africa is the most tropical of all continents. Due to Africa's latitudinal position, the continent is strongly influenced by the subtropical anticyclonic belts from both hemispheres. Consequently, it possesses extensive areas of dry climate both north and south of the Equator.

Unlike America or Eurasia, Africa lacks meridional mountain ranges that could act as climatic barriers or climate divides on a larger scale, so the continent, over large areas has a relatively undisturbed, almost ideal zonal, latitudinal arrangement of the main climatic zones. East Africa occupies a special climatic position. The altitude and the large water surfaces of the lakes cause non-zonal and smaller-scale circulation patterns. In contrast, the spatially limited mountainous countries of the Atlas in the north, the Ethiopian Highlands on the east side, and the Drakensberg along the southeast coast of Africa are of lesser importance in terms of climate. Greater influence, however, is exerted by the extensive plateaus, especially in South Africa, in addition to the large flat basins such as the Chad and Congo Basins in Western and Central Africa and the wide, little structured, and not very elevated desert areas of the Sahara, on which the volcanic mountainous massifs of the Hoggar and Tibesti are superadded.

Africa exhibits a hemispherical asymmetry in climate. Pronounced differences emerge between the climates of northern and southern hemispheric Africa, primarily due to the larger and much broader landmass north of the Equator compared to the southern region. Additionally, northern Africa lacks a true oceanic boundary, instead of being bordered by the Mediterranean to the north and the Eurasia landmass adjoining to the northeast. Thus, the continent can be divided based on the mean annual rainfall, which, due to its importance for agricultural use, is the most important climate element, into Wet and Dry Africa (Figs. 5.1 and 5.2). **Wet Africa** extends to the climatic dry line at about 1,000 mm annual



R. Baumhauer, *Physical Geography of Africa*, Geography of the Physical Environment,
https://doi.org/10.1007/978-3-662-71406-5_5

rainfall and thus includes Central Africa, the southern Sudan zone and the coast of West Africa, the southern and middle Highlands of Ethiopia, and the windward sides of the East African volcanic massifs with more than 1,800 mm annual rainfall and more than ten humid months or in Ethiopia six to nine humid months, parts of the highlands of southern East Africa as well as the Cape Fold Belt as "wet island" in dry South Africa. There, although three to five months have a higher temperature amplitude, the rainfall is still so high that reliable rainfed agriculture can be practiced. **Dry Africa**, on the other hand, receives only up to 500 mm annual rainfall in the wet-dry tropics or 250 mm in the dry subtropics, with precipitation in winter. This corresponds to the critical limit of the growing period. Rainfed agriculture is fundamentally still possible there, but livestock farming is practiced. At least six months of dry season and temperatures that are more strongly influenced seasonally lead to more frequent drought periods. Also, the **rainfall variability** increases significantly here, i.e., the reliability and amount of rainfall in space and time

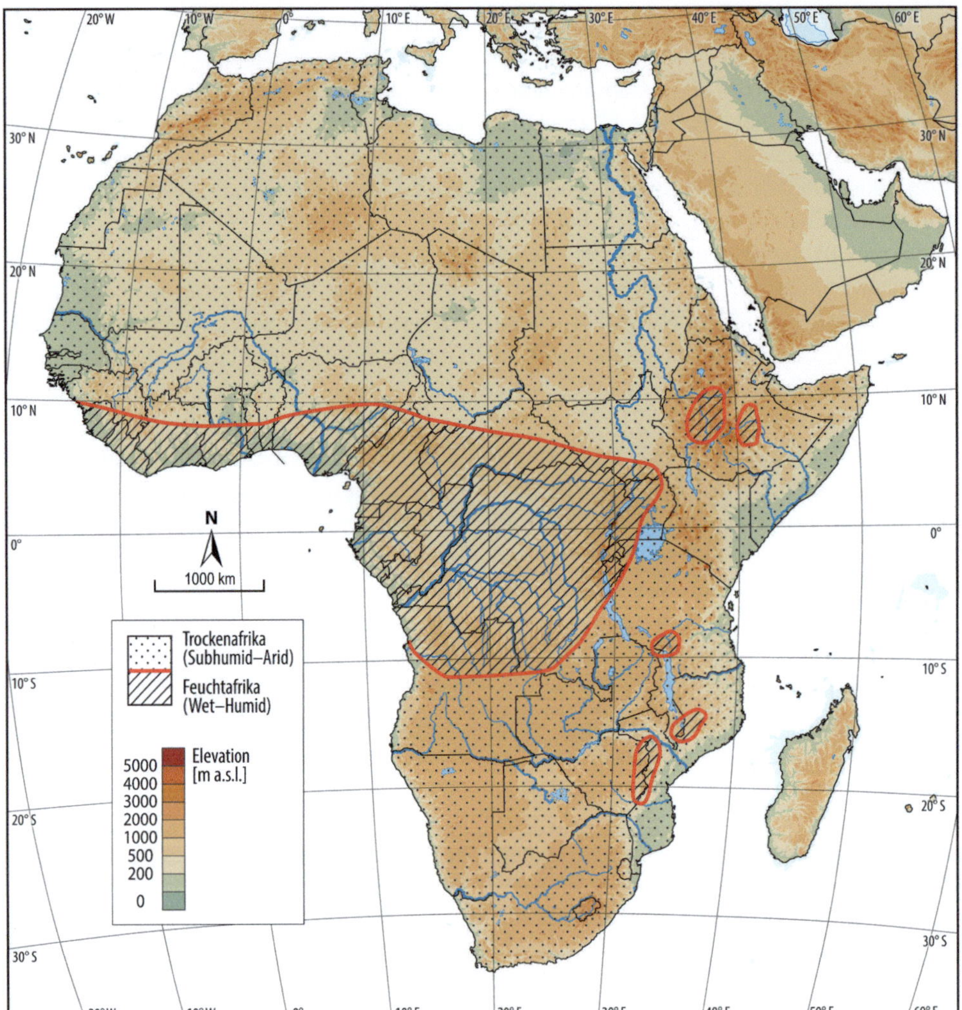


**Fig. 5.1** Wet and dry Africa (Feucht und Trockenafrika)

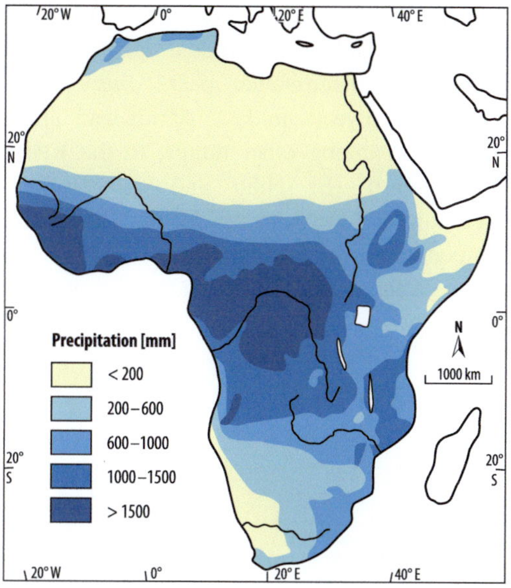


**Fig. 5.2** Distribution of mean annual precipitation of Africa

decreases. In North Hemisphere Africa, moving further north, the number of humid months decreases from four or five to two, then with less than two humid months and less than 100 mm annual rainfall, it transitions into the zone of dry tropical and dry subtropical (semi- and desert) climates of the Saharan region, which extends from the Atlantic to the Red Sea. With the same degree of aridity and only distinguished by the ratio of annual amplitude to daily amplitude of temperature, the subtropical "dry continent" of the Sahara follows northwards with the largest spatial extent of a climatic zone in Africa. A comparable transition from tropical (Kalahari) to subtropical (Namib) dry/semi- and desert climate takes place in southern central and southern Africa.

In contrast to the cool temperate climates, as we find them, for example, in Central Europe, there are no thermally influenced seasons like winter and summer in Africa, but the annual rainfall cycle dominates in the form of a typical alternation between rainy and dry seasons. Only in the Equatorial areas does the number of rainy days per week and month change. The characteristic climate rhythm is attributed to the seasonal north–south oscillation of planetary air masses, wind systems, and convergence zones (for example, the **Intertropical Convergence Zone ITCZ**, the trade winds, and the monsoons). This oscillation is driven by the changing position of the sun and the thermal contrasts between water/ocean surfaces and land masses. While the position of the ITCZ (which Flohn 1965, also referred to as the meteorological Equator, as it separates the northern hemisphere from the southern hemisphere atmospheric circulation) over the oceans changes only slightly during the year, it shifts considerably over the continent due to the stronger thermal reaction of the land mass.

The ITCZ corresponds, thermally and barometrically considered, to an Equatorial trough of low pressure, whose seasonal migration follows the sun's highest position in the respective hemisphere with about a one-month delay. In the Sahel region of West Africa, there are monsoonal rains from the Atlantic to Chad from May to October, with the main rainy season in July, August, and September. During this time, the **African Easterly Waves** (or African Waves) move along latitude-parallel bands, driven by the Guinea monsoon. The African Easterly Waves are wave disturbances in the tropical eastward flow, which occur on the Equatorward flank of the subtropical high-pressure areas and move from east to west. However, the alternation of dry **Westerly Waves** and moist Easterly Waves can account for more than 50 per cent of the annual precipitation variability in the Sahel in space and time. One consequence of the African Easterly Waves for the African climate is the breaking of the trade wind inversion, which leads to local precipitation, independent of the monsoon.

In southern Africa, the monsoonal rains fall in the southern summer, and in the Equatorial regions, there are two rainy seasons. They are determined by the position of the sun, which is the decisive factor for the division into rainy and dry seasons. However, local and regional topographical factors such as mountains or lakes also play a role (see also Figs. 5.3, 5.5 and 5.8).

In East Africa, the precipitation is unusually low compared to other intertropical areas, with annual precipitation totals between 500 and 2,000 mm, distributed over a twice-yearly zenithal rain. Indeje et al. (2000) pointed out precipitation peaks every 2–2.5, 3.5, and 5–6 years in East Africa and attributed these, among other things, to the **ENSO phenomenon** (El Niño and the Southern Oscillation 4.8–6 years) and **SSTs** (Sea Surface Temperatures) in the Equatorial Atlantic and Indian Ocean (see Chap. 9). In the east of South

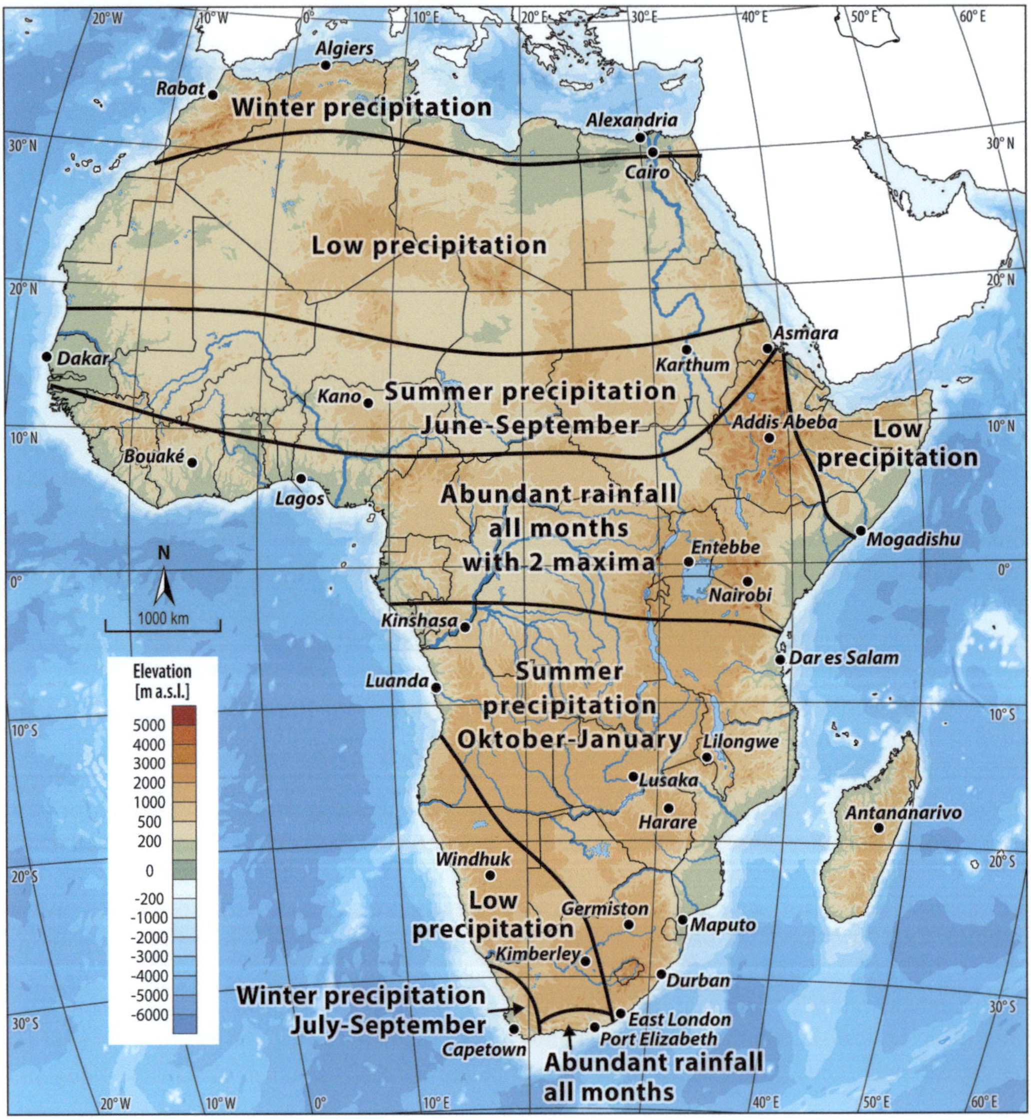


**Fig. 5.3** Seasonality of precipitation in Africa

Africa, the topography together with the effective ocean currents lead to significantly higher precipitation.

Much of Africa shows only a small annual range of temperature: Around 35 per cent of the continent's area has a range of less than 6 °C. Along the Equator, it is sometimes less than 3 °C. The diurnal range in temperature values is often great and is often more than the annual range. Over large areas, the diurnal range exceeds 15 °C and reaches 19 °C, e.g., in southeast Libya. Hot temperatures occur in large parts of the continent. Mean maximum values of more than 32 °C are ubiquitous. The highest temperatures are measured in the Sahara with mean temperatures of more than 38 °C (see Fig. 5.4). In contrast, mean minima below 5 °C are measured in only a few regions in the interior of southern Africa, in the northern Sahara and its border areas. Due to the high solar radiation, evaporation reaches high values. The maximum potential evaporation is in the Sahara at 2,500 mm/year, on average over the continent at 1,800 mm/year. Lower values are reached due to the more humid climate and denser cloud cover only around the Equator, so only 1,000 mm/year are measured at the Gulf of Guinea.

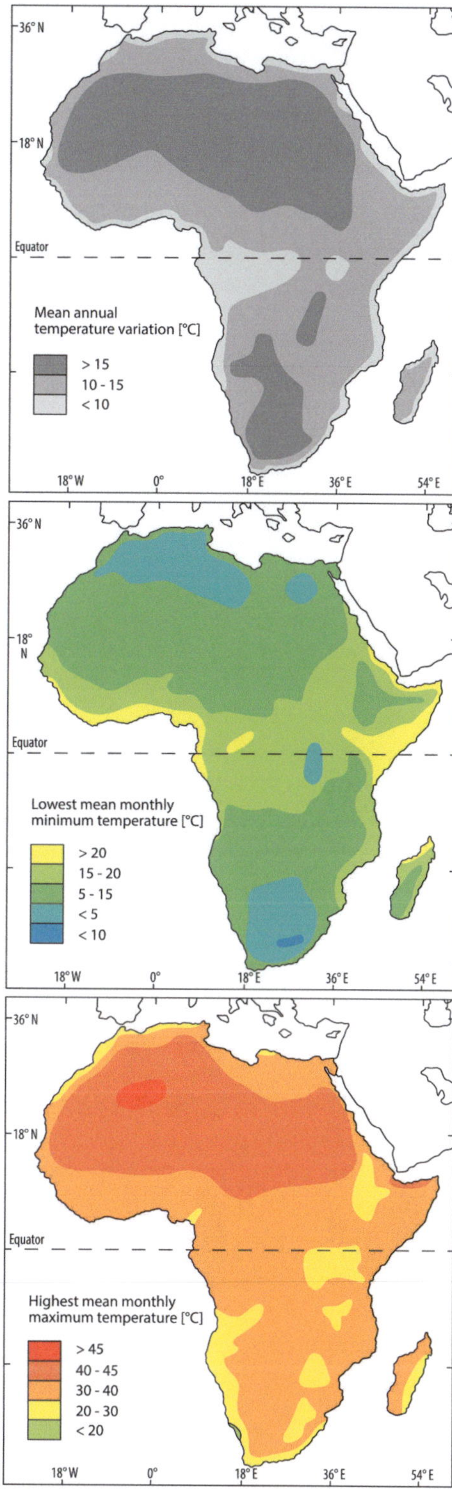


**Fig. 5.4** Temperature conditions in Africa

### Excursus: The Influence of Ocean Currents on Africa's Climate

Africa has a relatively straight coastline with few indentations. Measuring approximately 36,000 km in length, its coastline is comparable to that of Europe, despite Africa being roughly three times larger. The Gulf of Guinea on the Atlantic side, the Gulf of Aden on the Indian Ocean, and the Greater and Lesser Syrte on the Mediterranean coast are the only major bays. Due to the little indented coastline, the ocean currents around the continent are quite influential on the climate. The Canaries Current flows along the West African coast southwards to Cape Blanc beyond the Tropic of Cancer and

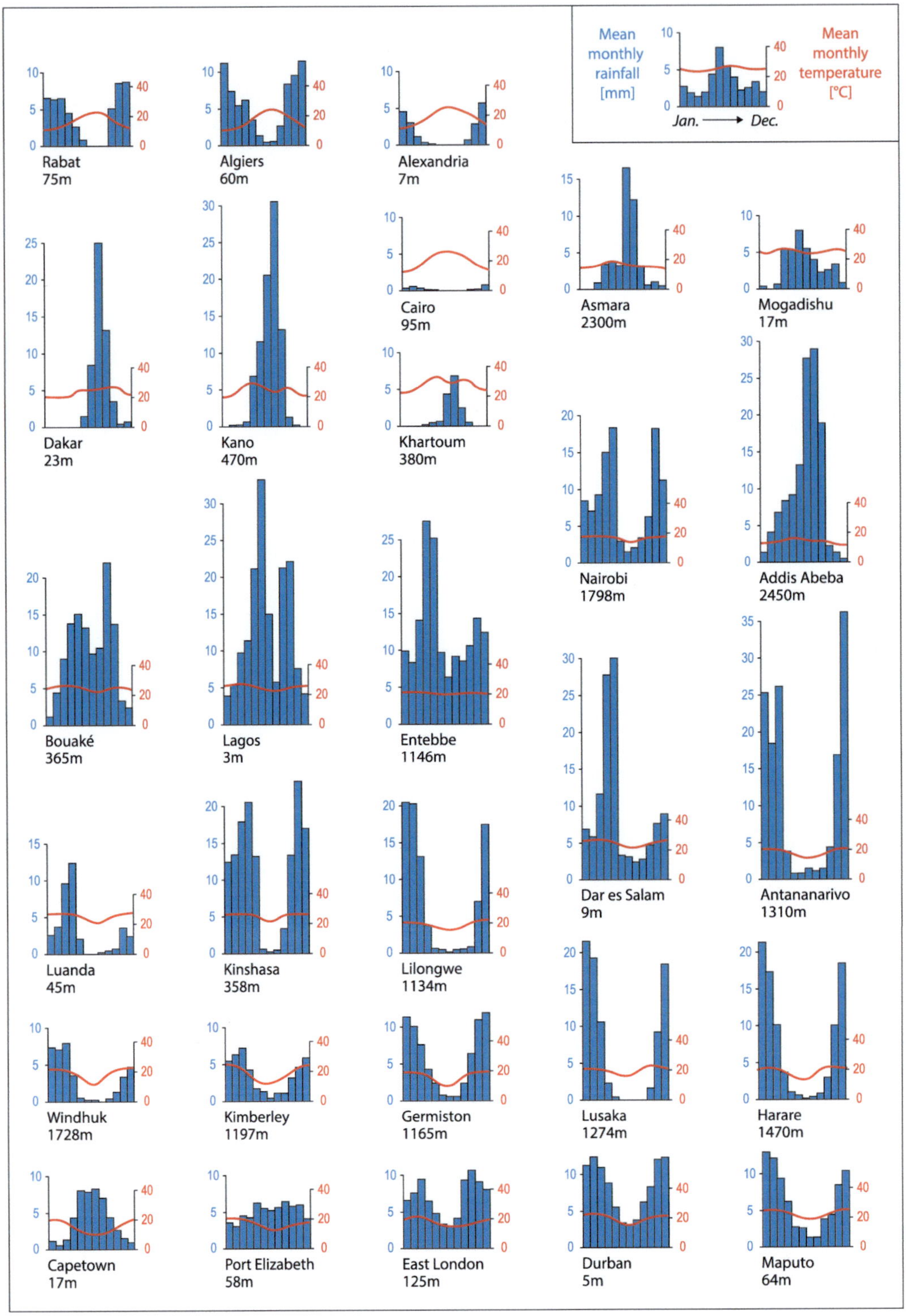


**Fig. 5.5** Mean annual precipitation and temperature in Africa

is between May and June about 6.5 °C colder than the "free Atlantic". The Benguela Current influences a very wide zone of 200 km between 23° and 31° S off the coast of Namibia and Angola. It is 3.5–5 °C colder than the South Atlantic. Thus, in Walvis Bay, the water temperature in December, which is in the midsummer, is only 11.5–14 °C. The third cold water upwelling center is located off the Somali Peninsula, the Horn of Africa. The temperatures in the Somali Current are between July and September, only 22.5 °C. The Canaries and Benguela Currents stabilize with their low water temperatures on the eastern sides of the subtropical belts of high pressure and have great significance for the high aridity of the coastal areas. Warm water centers are situated off West Africa, specifically off the coast of Liberia and in the Bay of Biafra. Sea surface temperatures in these areas consistently exceed 25 °C throughout the year. Additional warm water centers include the Mozambique Channel and the Red Sea.

## 5.1 General Atmospheric Circulation—The Tropical Circulation Over Africa

As a basis for understanding the climate and the climate dynamics, Fig. 5.6 shows the schematic representation of the earth's air pressure belts, and Fig. 5.8 shows the air masses, surface-wind systems, and convergence zones over Africa in January and July.

Figure 5.6 illustrates that the basic pattern of air movement is Equatorward, being from the northeast in the northern atmosphere and the southeast in the southern one. Converging air meets in the Equatorial zone, and here, air rises, moving back towards the poles at high levels only to descend in the vicinity of the two tropics. At the level of the tropics, subtropical belts of high-pressure are formed. The high-pressure

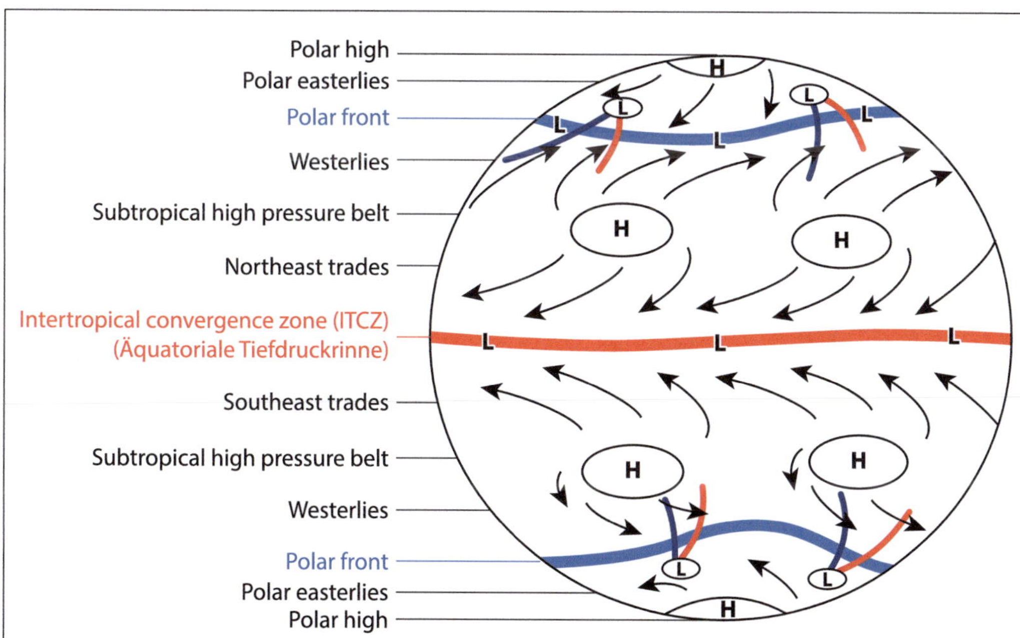


**Fig. 5.6** Schematic diagram of global surface- wind systems and pressures

belts, are dynamic and not thermal air pressure formations. This means they form in interaction with the adjacent westerly wind zone of the middle latitudes as dynamic, temporarily formed air masses on always comparable tracks. Essential components and decisive atmospheric action centers of the subtropical belt of high-pressure for Africa are on the Northern Hemisphere, the Azores High (and over Asia, the Ferrel's Pressure pattern) as well as on the Southern Hemisphere, the Saint Helena High over the Atlantic Ocean and the Indian or Mascarene High over the Indian Ocean. Between the subtropical high-pressure belts lie, separated by the trade wind circulation ("**Hadley Cell Circulation**"), the Intertropical Convergence Zone (ITCZ) or Equatorial trough of low-pressure. However, the position of the Hadley cells and the ITCZ varies with the seasons. This is related to the tilt of the Earth's axis and the annual circuit of the Sun made by the Earth, which means that the input of solar energy reaches a maximum in the northern hemisphere in June and in the southern hemisphere in December. The zones of subsiding air, the high-pressure cells (systems) associated with them, and the Equatorial westerlies in between travel north and south with the seasons. During the northern hemisphere summer, the Equatorial trough of low pressure migrates northward and low pressure extends throughout most of Africa north of the Equator. A marked reversal takes place in the southern hemisphere summer, when the subtropical high-pressure belt extends over most of northern Africa and the Equatorial trough of low pressure moves into the lower latitudes and Africa south of the Equator. The very northernmost part of the continent comes under the influence of eastward-moving mid-latitude low-pressure systems and westerlies. Thus, in January, the north-easterly trade wind (Harmattan) dominates most of North and West Africa, while north-east trades originating over the Indian Ocean (north-east monsoon) prevail over Eastern Africa. The subtropical highs are the control elements for the distribution of humid and arid climates and the seasonal change of precipitation in Africa. Large areas of the subtropical latitudes are characterized by almost year-round arid conditions due to sinking air masses in the global Hadley Cells (see Fig. 5.7).

The **Equatorial rainforest** is the only region where moisture absorption serves as a source for precipitation in the tropics bordering the Southern Hemisphere. The Southeast trade wind of the southern hemisphere reaches in the southern winter, unlike the southern summer, due to a low-lying trade wind inversion and the relief of the Great Escarpment hardly far into the interior of southern Africa. In the Equatorial area, moderate to strong zenith rains occur, which are caused by the convergence of the trade wind systems from both hemispheres in a low-pressure trough near the Equator, the Intertropical Convergence Zone.

The extreme northern and southern extremities of the continent are seldom affected by the poleward movements of Equatorial air masses. They are areas where winter rainfall predominates when cool, moist air and associated frontal depressions bring precipitation from higher latitudes.

In the **winter of the northern hemisphere** (see Fig. 5.8), the Azores High over North Africa forms a mostly continuous belt of high-pressure, which in the east connects to the central Asian cold high.

The high-pressure belt is particularly stabilized off the West African coast by the upwelling water of the Canary Current. However, it is interrupted by the warm Red Sea and its border areas. The high-pressure area reaches over North Africa only about 2,000 m upwards. Above it flows the (ectropic) westerlies with its passing low-pressure areas. Only Equatorward from 17° N does the subtropical high-pressure belt extend higher into the atmosphere, but then it is weakened near the surface by continental heat effects. Over southern Africa, a heat (thermal) low, with its center over the highlands ("Highveld") of Zambia and Zimbabwe, particularly dominates in the summer, with its center approximately in the northwest of the Kariba Dam. This forms a summertime low-pressure connection to the East African highlands and the Mozambique Channel, also referred to as the sITCZ. This second low-pressure trough, located further south

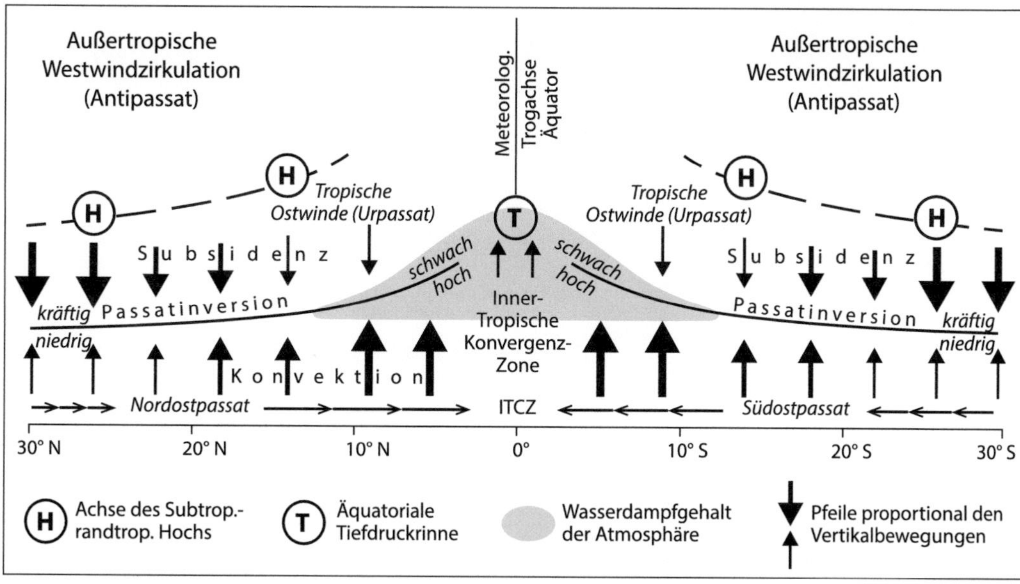


**Fig. 5.7** Schematic cross-section through the Hadley Cell Circulation system of the tropics. *Source* Weischet and Endlicher (2000, p. 256). Außertrop. Westwindzir kulation = Westerlies; Tropische Ostwinde = Equatorial Easterlies; Innertropische Konvergenzzone (ITCZ) = ITCZ; Nordost-Passate = Northeast Trades; Südost-Passate = Southeast trades; H = subtropical high-pressure belt (high); T = equatorial trough (low); Wasserdampfgehalt = water vapor; arrows are proportional to vertical movement

of the Intertropical convergence zone (ITCZ), significantly determines the regional weather events in southern Africa. The South Atlantic Saint Helena High, on the other hand, extends on the west side of Africa into the Congo Basin and forms the roughly SW-NE trending **Congo air mass boundary** ("Congo Air Boundary", CAB) over the continent.

The actual intertropical low-pressure trough (ITCZ) is located over the Atlantic at 3° N, reaches the African coast near Monrovia in Liberia, and then runs north of the Guinea coast via Ivory Coast, Nigeria, and Cameroon to the northeast of the Democratic Republic of Congo and further over the western Rift into southern Malawi. West of Madagascar, it touches the Mozambique Channel. The northeast trade wind emanating from the Azores High is limited in its maritime variant to the African west coast in Morocco and Mauritania. It can converge with the continental variant, which is very dry due to its origin in the Central Asian high-pressure area and the long journey over the continental Sahara. This area of trade wind convergence extends over western Africa near the coasts of Senegal, Sierra Leone, and northern Liberia. However, North Africa is mainly subject to the northeast trade wind in its dry Harmattan variant. In eastern Africa, it can cross the Equator and flow into the heat lows over Zambia and the Mozambique Channel.

The maritime southeast trades of the Saint Helena High reach the meteorological Equator, the ITCZ, over the Atlantic west of West Africa, is deflected to the right after crossing the Equator due to the change in the direction of the Coriolis effect and thus becomes the southwest monsoon. The semicircular arrangement of heat lows in their weakened form over Sudan and Central Africa and a stronger form over the East African highlands and especially over Zambia leads to a fan-like widening of the monsoon flow and bending from a southwestern to a northwestern direction in southern Africa. The relative humidity of the southwest monsoon, which is already relatively enriched with water vapor due to its maritime origin, is further increased to 80–85 per cent by the influence of the rain forest. The southeast

**Fig. 5.8** Mean monthly atmospheric pressure and prevailing surface winds for January and July. Pressure units are millibars reduced to sea level. Air masses, currents, and convergence zones over Africa in January and July

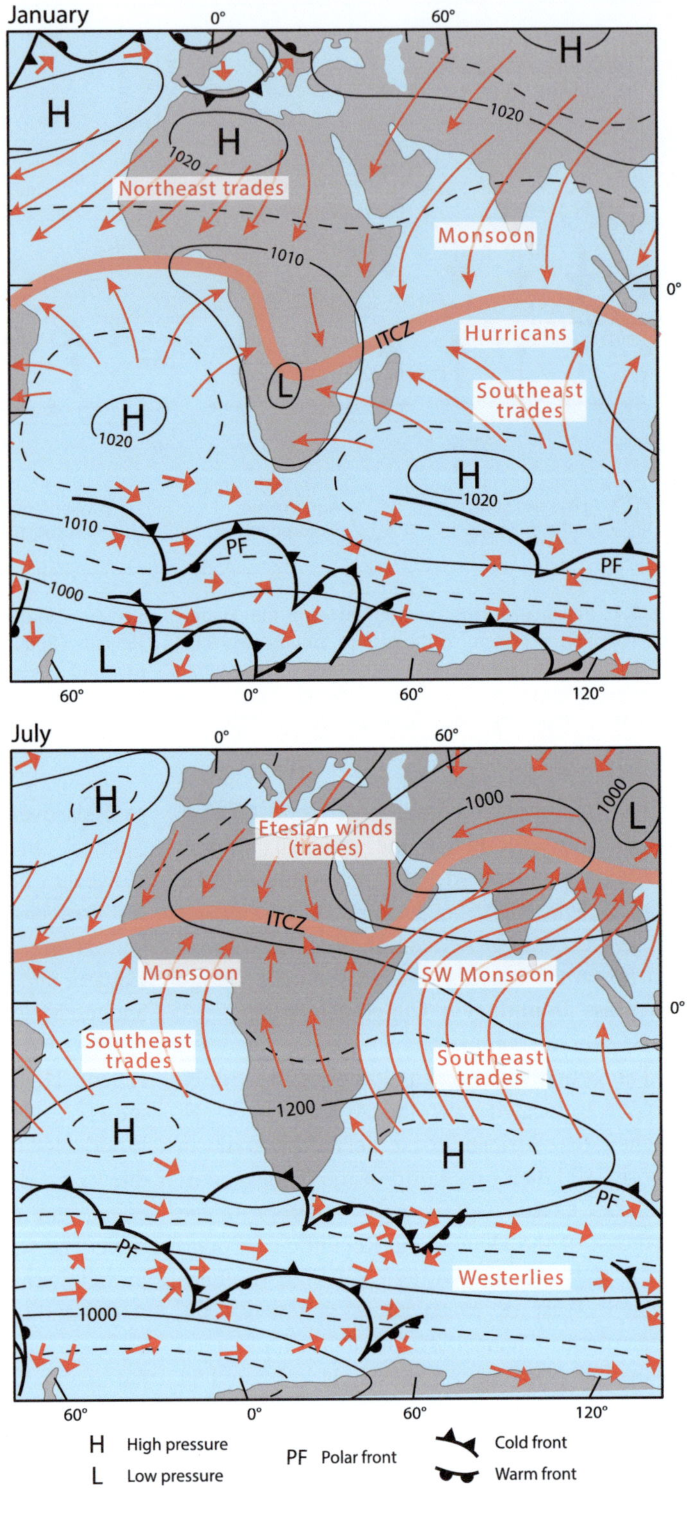

trade wind of the Mascarene High forks east of Madagascar into a northern one that flows directly into the ITCZ, and a southern one that bypasses Madagascar in the south and penetrates the continent in the Zambezi Valley. There, it is deflected anticyclonically in southwestern directions towards the interoceanic confluence over the western highlands.

**Excursus: METEOSAT**

METEOSAT, short for Meteorological Satellite, is a constellation of geosynchronous European weather satellites. These satellites were developed in close cooperation with the European Space Agency (ESA) and are operated by the European organization EUMETSAT. Since the commissioning of the first satellite in 1977, METEOSAT has provided weather information for the regions of the Earth around the zero meridian. The geosynchronous position used at 0° at a geographical longitude and an altitude of approximately 36,000 km above the Equator is optimal for weather observation over Africa, the eastern Atlantic, and southern Europe. Since the beginning of 2004, the first METEOSAT satellite of the second generation (short MSG-1 for Meteosat Second Generation) has been operational. The satellites, which replaced the second-generation MSG in 2023, bear the designation METEOSAT Third Generation (MTG).

The January image in Fig. 5.9 shows the characteristic atmospheric circulation in northern midwinter and southern midsummer. In the polar region of the northern hemisphere, there is a polar night, and the sun does not rise above the horizon at noon from 67° N poleward. In contrast, the southern hemisphere is brightly illuminated up to the South Pole. With the extreme solar zenith corresponding to the season near the Tropic of Capricorn, all global wind and pressure belts and the circulation pattern associated with them are shifted far southward. The phenomena characterizing this over Africa and its border areas are rainy winter weather in the latitudinal subtropics from Madeira to parts of the Persian Gulf, in the high altitudes of the mountains and with snowfall. Dry season in the northern tropics and also in Equatorial Africa. Far-reaching "meander arcs" of the Intertropical convergence zone (ITCZ).

Over the northern part of Africa, the subtropical belt of high-pressure is strongly developed and extends particularly far into the wet tropics, a situation that is characteristic of the season. January is the driest month in (always wet) Equatorial Africa. The present image shows one of the rare weather situations in which the air from the Sahara to the southern Congo Basin is so low in water vapor that not only the transition from the desert into the thorn tree and desert grass and tall grass savannas and the savanna woodland is clearly visible, but in the Congo Basin itself the course of the Congo and the lakes in the Rift Valley are clearly visible. The subtropical high-pressure belt of the northern hemisphere extends in January due to the shift of the solar position only up to the coastal section of North Africa.

The position of the ITCZ with its intense convective weather events shifts, as the climatological rule says, seasonally with a slight delay compared to the position of the sun. The southern position near the tropics therefore corresponds to the season. The wide meander swing of the ITCZ comes about because it always oscillates over the Atlantic, even in the southern summer in the area between 0° and 10° N, and does not significantly exceed the Equator on average.

In the summer of the northern hemisphere, the atmospheric circulation over Africa fundamentally differs from that in the northern winter. The influence of the Azores High remains limited to the Atlantic and the Mediterranean. Surface high-pressure cells are, therefore, only west of the Moroccan coast and north of the coast of Algeria, Libya, and Egypt in the form of a ridge. The conditions over North Africa are different: Over the

**Fig. 5.9** Image from 01/21/2020, time: 12:00 UTC, spectral range: visible light. *Source* EUMETSAT

Sahara, a surface heat low forms, into which the continental dominated the wind-flow pattern of the Harmattan from the northeast, the southwest monsoon from the southwest, and the deflected maritime northeast trades from the northwest flow. The near-surface atmosphere is desert dry with just 15 per cent relative humidity. The ITCZ lies over the Atlantic at 12° N south of the Cape Verde Islands and reaches the coast at the latitude of Dakar but swings out over the continent to about 15° N. Over southern Algeria, Chad, and Sudan, it extends to the Red Sea at Port Sudan, and there is a direct connection to the surface low over the Arabian Peninsula to the monsoon low in the Middle East. The ITCZ over North Africa is superimposed in height by the subtropical high-pressure belt. This is characterized by a strongly pronounced subsidence movement, which is due to the influence of the subtropical east jet. As a result of the stable high altitude over High Asia, the air pressure difference between it and the ITCZ in the upper troposphere is compressed to a relatively short distance. This results in a permanent self-reinforcement of the dynamic northern hemisphere subtropical high-pressure cells. This is responsible for the extreme dryness of the Sahara.

In the monsoon of West Africa often, mesoscale convective regional systems with a diameter of 10–100 km develop, which can produce intense heavy rainfall of 50–300 mm per day. This pattern is rooted in the African Easterly Waves: In the summer monsoon circulation (April to September), warm, moist tropical air meets even warmer air from the Sahara at about 15° N. The result is a thermally induced easterly wind increasing with altitude and a thermal equilibrium level at about 4,000 m altitude, with the African Easterly Jet as the wind maximum. This jet is relatively unstable and is stimulated to meander by mountain ranges and near-surface heat islands, the African Easterly Waves, which can trigger tropical cyclones over the Atlantic. One consequence of the African Easterly Waves for the African climate is the breaking of the trades, which is accompanied by abundant local rainfall independent of the monsoon.

**Excursus: Monsoon**

Monsoons are winds that change their main flow (wind) direction by at least 120° in a seasonal cycle, which is accompanied by a significant change in weather patterns. They are fundamentally different from the trade winds. They are triggered by shallow Ferrel's heat lows, which usually form over the dry highlands during the summer half-years of the respective hemisphere. They can arise from redirected trade winds of the opposite hemisphere or blow directly from the trade winds into the monsoon low. The most famous monsoon is known as the Indian summer monsoon and is accompanied by heavy rainfalls in south and southeastern Asia. In the summer of the northern hemisphere, an Asiatic low is produced by the intense summer heating of the landscape. The low draw-in warm, humid air masses from the Indian Ocean, which move from East Africa as a continental-offshore southwest monsoon, northward, brush the Horn of Africa and pass over India, Indochina, and China. The southwest monsoon of West Africa, which influences the Guinea coast countries and Sudan states far into the Sahel zone, arises in the southern summer from the southeast trades of the southern hemisphere, blowing out of the Mascarene high area, which is redirected into an onshore southwest monsoon when crossing the Equator.

The Saint Helena High of the South Atlantic is pushed far north in the southern hemisphere's winter. Its influence is further enhanced by the summer upwelling of cold deep water off the Republic of Côte d'Ivoire. Southern hemisphere high-pressure areas are connected close to the ground by a high-pressure ridge over southern Africa. The sea winds emanating from the Saint Helena High cannot surpass the Great Escarpment of Namibia. In the east, these winds are superimposed in height by the continental descending southeast trades of the Indian Ocean and by autochthonously dry air masses originating from the Ground High of the Kalahari. Under certain weather conditions, the latter can form in the Namib Foehn, a type of dry, relatively warm downslope wind in the lee of the Great Escarpment. This has the interesting effect that in the coastal stations of Walvis Bay or Swakopmund, the highest temperatures are often measured in winter, as the dry, in the highlands, around 20 °C warm east winds can heat up to almost 40 °C during dry adiabatic descent. In contrast, in the southern summer, the strong heating of the continental interior leads to a shift of the Congo Air Boundary (CAB) with strong thermal convection, with the coast coming under mild oceanic influence. The interoceanic confluence, which lies all year round over the abrupt drop of the large escarpment from Angola to Namibia, is not very weather-effective.

In the summer hemisphere in Fig. 5.10, the high-pressure belt encompasses the Mediterranean, reaching nearly 45° N and stretches eastward along the Tropic of Cancer to the eastern end of the Arabian Peninsula. Its center lies over the northern Sahara, the southern Mediterranean, and the Red Sea at about 30° N. The mountain edges of Ethiopia (also of Yemen and Oman) are marked against the respective offshore warm seas by closed rows of cumulus clouds because of the landward and mountain-upward wind.

On the southern hemisphere, the corresponding high-pressure area with its axis is set at approximately 20° S. The Equatorward edge extends over Africa to the Lunda Ridge. The Intertropical Convergence Zone lies in West Africa, according to the season, over Sudan and the Guinea coast countries. Directly on the coast of southeastern Africa, there is a solid front of

**Fig. 5.10** Image from 07/14/2021, 12:00 UTC, spectral range: visible light. *Source* EUMETSAT

mighty convection clouds. They are the result of air masses transported landward by the southeast trade winds, charged with moisture over the warm Agulhas and Mozambique Current, which meet the Great Escarpment there, which is over 2,500 m high.

In the transitional seasons, there is only a rapid shift of the Equatorial low from the northern to the southern hemisphere over East Africa after the equinoxes. In the western and central parts of the continent, the low remains in the northern hemisphere all year round. No low forms over the Congo Basin, which can be attributed to the stabilizing effect of the sea and its continuation in the form of the rain forest into the interior of the continent. In West Africa, too, the surface low does not reach the coast but remains static north of the rainforest strip. In addition, the ITCZ of the middle troposphere, responsible for the precipitation, deviates much less far to the respective summer hemisphere. It stays for four months in its extreme summer situation and only influences the Equatorial area for two months during the transition seasons, that is, despite the sun's highest position twice, the tropical Equatorial zone represents a rain-poor zone.

Since the recording (Fig. 5.11) was made at the time of the equinox, the solar radiation conditions on the northern and southern hemispheres are the same. The ITCZ remains 5–10° south of the Equator. While its poleward extent is limited, it merges with a broad transition zone equatorward, where powerful thunderclouds are present. Over the Gulf of Guinea and the Atlantic, the ITCZ is remarkably far south, only 100–200 km compared to normally about 500 km north of the Equator. This can be seen as

**Fig. 5.11** Recording from 03/18/2014, 12:00 UTC, spectral range: visible light. *Source* EUMETSAT

a seasonal effect: On the northern hemisphere, a huge closed high-pressure area is formed on both sides of the Tropic of Cancer, extending from the Persian Gulf across North Africa and the Atlantic to the northeast coast of South America. In Europe, a wedge of a high extends over a large part of Western, Central, and Eastern Europe.

In the southern hemisphere, the corresponding subtropical high is relatively weak and spatially divided into two centers. The eastern high-pressure center is indicated by the cloud-free area in the north of South Africa, with an offshoot to Namibia. The actual center of the southern hemisphere subtropical high lies over the southern South Atlantic in the sea area between the islands of St. Helena and Ascension, marked by the characteristic stratocumulus clouds. They occur with great regularity in this region but are weak currently.

The recording of Fig. 5.12 clearly shows the equatorial proximity of the southern hemisphere's subtropical high-pressure belt. It is characteristic of this season, and it corresponds to the seasonal shift. Especially over the oceans, the axis of the subtropical belt of high-pressure lies a few latitudes further Equatorward in the winter of the respective hemisphere than in the summer half-year.

Over mainland Africa, the convective clouds with mighty thunderclouds in the west of Central Africa only start at the level of the Congo's mouth. In the east, even the entire southern Congo Basin up to about the latitude of Burundi has almost clear weather. On the polar flank of the high-pressure belt, the cyclonic westerlies from the Subantarctic have prevailed far into the subtropics. This is particularly beautiful to see in the sea area off the South African east coast–south-southeast of the southern tip of

**Fig. 5.12** Recording from 09/10/2021, 12.00 (UTC), spectral range: visible light. *Source* EUMETSAT

Madagascar, at about 30° S, is the center of an extratropical cyclone. The cold front runs in a northward convex arc close south of the southern tip of Madagascar towards the east coast of Mozambique and South Africa. It is followed by the pearl necklace-like arranged cumulonimbi with so-called open cells and cloud streets.

On the southwestern side of the continent, the typical coastal, but separated from the land by a narrow fog-free strip, a fog patch can be seen in central Namib. It does not reach the mainland due to the typical offshore mountain wind situation of this season.

As a result of the extreme expansion of the southern hemisphere's subtropical high-pressure belt into the tropics near the Equator, the subsequent zone with tropical-convective weather events is particularly far poleward. A notable example of this can be seen in the weather conditions prevailing over Eritrea and eastern Ethiopia. Thus, the Highlands of Ethiopia towards the northeast and the northern edge of the Horn of Africa against the Gulf of Aden are marked by small but mighty thunderclouds. Compared to the situation described for August, in which dryness prevailed throughout East Africa under the influence of the northern outflow on the western flank of the monsoon low, a change in weather has thus occurred with the seasonally corresponding decay of the monsoon low. However, the dry season currently prevails in the East African highlands south of the Equator.

Over Central Africa, the Guinea coast countries and the Equatorial Tropical Atlantic, the Intertropical Convergence Zone (ITCZ) would be expected at this time of year. However, the satellite image shows a synoptic tripartition in the atmospheric events: An actual convergence area with corresponding cumulonimbus clusters from the western Congo Basin to Nigeria, northeast trades without rain clouds from the Volta reservoir in Ghana to the west of the Rep. of Côte d'Ivoire and an adjacent cyclone, which extends from Liberia to Senegal and far into the adjacent Atlantic.

In the Northern Hemisphere, the area of influence of the subtropical high-pressure belt is less extensive than in the Southern Hemisphere, and it lies further poleward. According to the season, anticyclonic weather conditions prevail over the northern edge of Africa and the southern part of the Mediterranean area up to about 37° N.

## 5.2 Climate Regionalization

Due to the unique symmetrical location of the African continent to the Equator, which results in a model-like latitudinal parallel arrangement of the climate and vegetation zones (see Chap. 8) of the tropics and the subtropical border areas, it is possible to divide Africa into **eight major climatic zones** (or regions, Troll and Paffen 1964), related essentially to precipitation and temperature characteristics. The presentation of the climatic zones thus deviates for pragmatic reasons from the previous consideration of the physical geography based on the major regions of the continent.

Africa can be divided into distinct climatic zones based on precipitation, temperature, different weather patterns, continentality, and altitude. One such zone is the **wet equatorial climate**, which dominates a vast area spanning the Equator in the Congo Basin and along the Upper Guinea coast of West and Central Africa. It is characterized by rain in all months and uniformly high temperatures throughout the year. Surrounding this climatic type is the **wet-dry tropical climate**, which has three to five dry months and a larger temperature range than in the wet equatorial climate. Precipitation and temperatures, however, remain high. This climate occurs in the Sudan area of West and Central Africa and merges into the zone of the **dry tropical climate**, which has low rainfall for at least six months. Rainfall amounts are lower and more variable than in the two previous zones, temperatures show a larger seasonal swing, and drought, and famine are recurrent problems in the Sahel of West Africa. The **dry subtropical climate** or **desert climate**, which is characteristic of the Sahara and the Namib, is the most extensive climatic type of Africa. In addition, there are the hygrically differentiated highland climates of East Africa as well as the **dry subtropical climatic zone with precipitation in summer**. They are found in the interior plateaus and basins of Southern and Central Africa, including much of the Kalahari semi-desert, with rainfall concentrated in about three summer months. Finally, at the northern and southern extremities of the continent, the Maghreb and the Mediterranean region and the Cape region of South Africa, the dry subtropical climatic zone with precipitation in winter, which is as well often described as having a **Mediterranean type of climate** (see Fig. 1.3).

The wet **equatorial climate** is characterized by a mean annual precipitation of more than 1,500 mm per year, spread over more than ten humid months. These climatic conditions are sufficient for the formation of the tropical rainforest. It is also referred to as Equatorial or rainforest climate and predominates in Central Africa between 3° N and 4° S and along West Africa's Guinea coast up to 8° N. However, the center of the Congo Basin receives less precipitation compared to the coastal areas (see below), with 1,500–2,000 mm per year. In Kinshasa, the capital of D.R. Kongo, the annual rainfall even drops below 1,400 mm. The center of the basin at 20° E and the western slope of the Rift Valley in the east are the wettest, with sometimes more than 2,200 mm of annual rainfall. After the Amazon Basin, Africa has the largest extent of a closed rainforest area (see Chap. 8). The high moisture content of this hygromorphic forest, whose surface temperature is significantly lower compared to the surrounding dry areas, is particularly important for the regional climate and thus prevents excessive heating of the earth's surface. In addition, the rainforest ensures the replenishment of water vapor for the air masses crossing it. The characteristic large-droplet tropical rainfall falls in the form of convective thunderstorms. In the months with the maximum rainfall after the equinoxes in April and May, as well as October and November, it rains at least every third day. During the two relatively dry periods, convective showers with 50–100 mm of rainfall occur occasionally, about every fourth day. Due to the convective rainfall, the inner Congo Basin forms a climatological exception. However, climate, weather, and weather conditions in the Congo Basin and its border areas are characterized by spatial–temporal variability. Thus, rainfall can be delayed by several weeks even in the wet Equatorial rain climate and in the alternating humid border areas in the change between rainy and dry season, large delays can also occur or, although quite rarely, completely fail. Thunderstorms occur on average one hundred days per year, most frequently with 180–200 days in the highlands east of the Democratic Republic of Congo. Here, the region of Bukavu with an average of 221 thunderstorm days, is even the most thunderstorm-rich area on Earth.

In the wet tropical coastal areas of West Africa, under the influence of the ITCZ, there are two maxima with excessive rainfall each in March and April as well as September and October, while in June/July and December/January, the amount of rainfall and the frequency of rainfall slightly decrease. The southwest monsoon meets the coast at a right angle in the coastal area between Guinea and Liberia. Rainfall amounts at the Guinea Ridge (2,000 m) in Sierra Leone are, e.g., in Conakry, over 4,000 mm per year. Another pronounced rainfall area is in the Bay of Biafra in southern Nigeria, Cameroon, and Gabon. There, due to the topographical situation of the concave coastal shape and the coastal mountains, record rainfall values are measured: In Debundscha in the mean up to 9,895 mm, and in San Antonio de Ureca on Fernando Poo, even up to 10,450 mm of annual rainfall. The rainfall decreases with the increase in continentality.

A notable rainfall anomaly occurs along the coastal section extending from southern Ghana via Togo to Benin. The so-called Dahomey Gap consists of a dry area sandwiched between more humid climates to the east and west, with the greatest intensity of dryness along the coast. There are less than 1,000 mm of rainfall, for example, in Accra 787 mm per year. The coastal trend and upwelling of cool water from the Benguela Current into the South Equatorial Current lead to a seasonal subsidence inversion, particularly in the Gulf of Guinea, where it extends up to 30° W, thus reducing convection. This results in a "Little Dry Season" from July to September, precisely during the main monsoon season in West Africa, along this stretch of coast.

The wet Equatorial zone of Africa has a mean annual temperature ranging from 24 to 26 °C. Typical of the tropics, this region experiences a diurnal climate, characterized by a minimal annual temperature range of only 1–2 °C, while the daily amplitude is 10 °C or more. The seasonal temperature change so typical for the ectropes is replaced in the equatorial climate by a temperature change tied to the times of day: cool and foggy morning hours (mean daily minimum: 18–20 °C), warm mornings, hot and humid midday and afternoon hours (mean daily maximum: 29–30 °C), thundershowers in the late afternoon or early evening hours and cooling down to 18–20 °C during the night hours—a constantly similar sequence almost every day of the year. Another characteristic of the wet equatorial climate, especially on the coasts, is the high sultriness. With a mean daily humidity of 85 per cent and mean daily temperatures of 25 °C, there is permanent humidity in coastal cities like Lagos, Douala, and Libreville. A major control on the annual amplitude of the temperature is the altitude. Thus, in the city of Bamenda, located at an altitude of 1,520 m on the highland of the same name in Northwest Cameroon, a mean annual temperature of 19.5 °C and an annual amplitude of 3 °C are measured, compared to 1.5 and 26.4 °C on the coast in Douala.

With the distance from the Equator, the number of humid months and the mean amount of precipitation decrease from nine humid months in a transition area to six to seven humid months with annual precipitation between 1,200 and 1,600 mm north of 3° N and south of 4° S. This is accompanied by an accentuation of the rainy seasons due to the seasonal shift of the ITCZ in the northern hemisphere in June and September, in the southern hemisphere in April and October. The rainforest belt in this transition area to the **wet-dry tropics** is replaced by Savanna woodland, paired with gallery forests of the rainforest type. With overall rising temperatures, a pronounced seasonal alternation between dry and rainy seasons begins in this zone. These wet-dry tropics extend in West Africa up to 14° N and can be divided into a southern Sudan zone with a short dry season and a northern zone with a longer dry season. The southern Sudan zone receives ample rainfall, with 1,000–1,500 mm of precipitation per year, primarily during the rainy season from May to September. In contrast, the northern part of the Sudan zone experiences more moderate precipitation, ranging from 500 to 1,000 mm. This northern area, where rain-fed agriculture is no longer generally possible, extends in a narrow strip around 13° N in the western part of West Africa via Burkina Faso and southwest Niger at 11–12° N and Chad at 10° N up to 9° N in the Bahr al-Ghazal of western Sudan. Especially the higher variability of precipitation in space and time can lead to serious limitations of agricultural use here (see Chap. 7). In Cameroon, north of the Adamawa Mountains and the northern parts of the Central African Republic, the dry northeast trade wind (Harmattan) from the Sahara influences the climatic conditions in the winter half-year: Bangui (4° 22′ N) has ten humid months at a mean annual temperature of 26.1 °C. The rainy season lasts from February/March to November/December. In the northeast of the country near Birao (10° 17′ N), the mean annual temperature is 32 °C, and there are only four to five humid months. The rainy season is limited to April/May to October, with a high interannual variability of the beginning and end. Similarly, the mean annual temperature shows a larger annual swing (Bangui 4.5 °C/Birao 6.5 °C), while the number of rainy days (130–72) and the amount of precipitation (1,560–860 mm) decrease. The highest daily temperatures and the greatest humidity are reached in the transitional period between the dry and rainy seasons. This climate is often referred to as the Sudan climate in West and Central Africa, further north and with increasing aridity, with annual precipitation amounts below 500 mm, as the Sahel climate.

In the southern part of the Congo Basin and the Katanga or Shaba Plateau belonging to the Lunda Ridge in the southeasternmost part of DR Congo, the climatic conditions change comparably, but not in such a model-like latitudinal way and manner as in the northern Congo Basin and its northern border areas.

The **wet-dry tropical climate of the Sahel** merges the Sudan zone to the north as a 3–4° wide transition area to the Sahara in the north. The Sahel region spans across Africa, from the Atlantic to the Red Sea, covering a vast area. In the west, it also encompasses the Cabo Verde Islands, located off the coast of southern Mauritania and northern Senegal, at a latitude of 15–16° N. It expands in the highly continental southern Chad and northern Cameroon equatorward to 12–13° N and extends over Kordofan to 16° N in northern Ethiopia. The mean daily maximums rise in northern Cameroon on the southern shore of Lake Chad in the warmest months of April and May to over 40 °C, the mean minimums do not drop below 12–13 °C in the coolest months of January and February. From south to north, the number of humid months first decreases from four to two, then transitions into the dry subtropical zone of the southern Sahara. With a mean annual rainfall of 150–500 mm, which falls as zenith rain in the summer rainy season, the Sahel is already beyond the critical limit of the growing period of rainfed agriculture of 500 mm per year. For agricultural use, the variable rainfall amounts are much more important than the mean amount. Together with the high intensity, the heavy rainfalls promote water erosion and intense surface runoff. For example, in July 1963 in Niamey (Rep. of Niger), about 200 mm of precipitation fell within one hour (see Figs. 5.13

and 5.14). The intense rainfall in this region provides limited plant-available water, making it relatively ineffective for agricultural purposes. With a decreasing amount of annual precipitation, its variability also increases. The extremely complex interactions often lead to droughts in the Sahel, the effects of which can be catastrophically amplified by anthropogenic influence. Comparable problem situations are also found in the "Sahel" of the southern hemisphere.

Low precipitation levels, extreme dryness of the air, high interior diurnal and seasonal temperature ranges and low cloud cover are the climatic characteristics of the **dry-subtropical zone**. The Sahara is the largest hot desert on our planet, and with an average of nine to eleven hours of sunshine per day, it is also one of the sunniest areas on Earth. This is also reflected in the thermal conditions. The mean daily maximums in summer are everywhere above 40 °C; and the absolute maximums ubiquitously exceed 50 °C. In June 1913 and 1922, the highest temperatures so far were measured in Libyan El Aziziyah at 56.7 and 58.0 °C. On the other hand, frost is widespread in the central and northern Sahara during the winter half-year. Over large areas, the seasonal range shows an elevated temperature amplitude.

At the northern extremity of the Sahara is the **zone of dry subtropical climate with precipitation in winter** from drifting cyclones of equatorward-extending fronts of depression from the westerlies. On the southern extremity of the Sahara, however, monsoonal convective summer rains fall. Both rain regimes only marginally influence the desert. They are not very productive and extremely variable in space and time. No desert on Earth surpasses the Sahara's continental-scale hyperaridity, which is further intensified in the Eastern Sahara by the Easterly (East) Jet, a wind pattern that intensifies the region's dryness. The Eastern Sahara is "the desert in the desert" and has a mean annual precipitation of 1–5 mm (for example, in the Kufra oases of Libya), one of the driest areas on Earth. Precipitation usually only occurs every three to five 5 years. The area remains year-round north of the ITCZ and under the influence of the northeast trades, referred to here as Harmattan. In other areas of the Sahara, annual precipitation of 5–50 mm is to be expected. There, the number of precipitation days is also higher, often due to passages of ectropic cold fronts, and can advance far to the south. For example, on our expeditions in 1981, 1982, and 1984, two to six cold front passages were observed

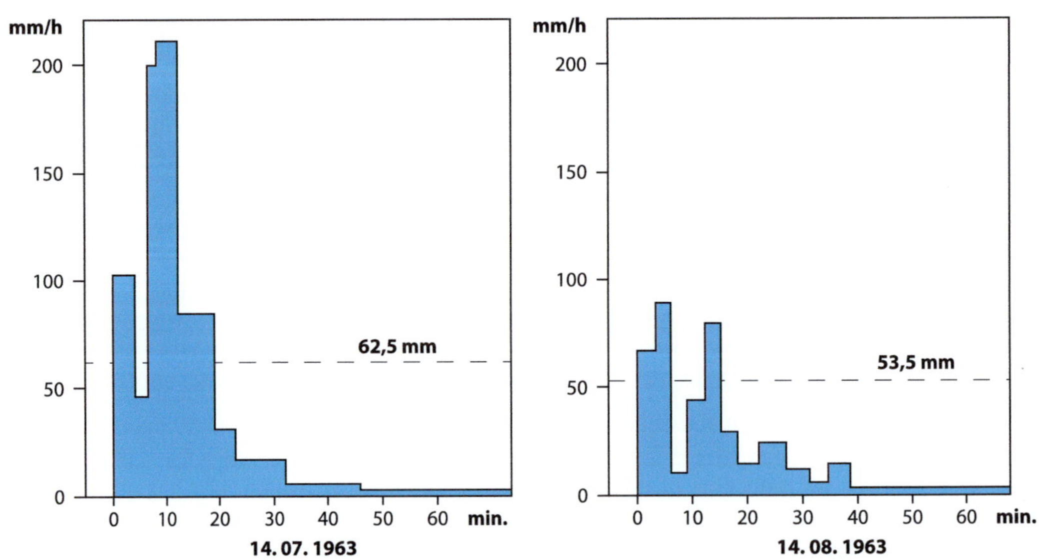


**Fig. 5.13** Precipitation record of two events in Niamey, Republic of Niger. *Source* Herbaud (1966)

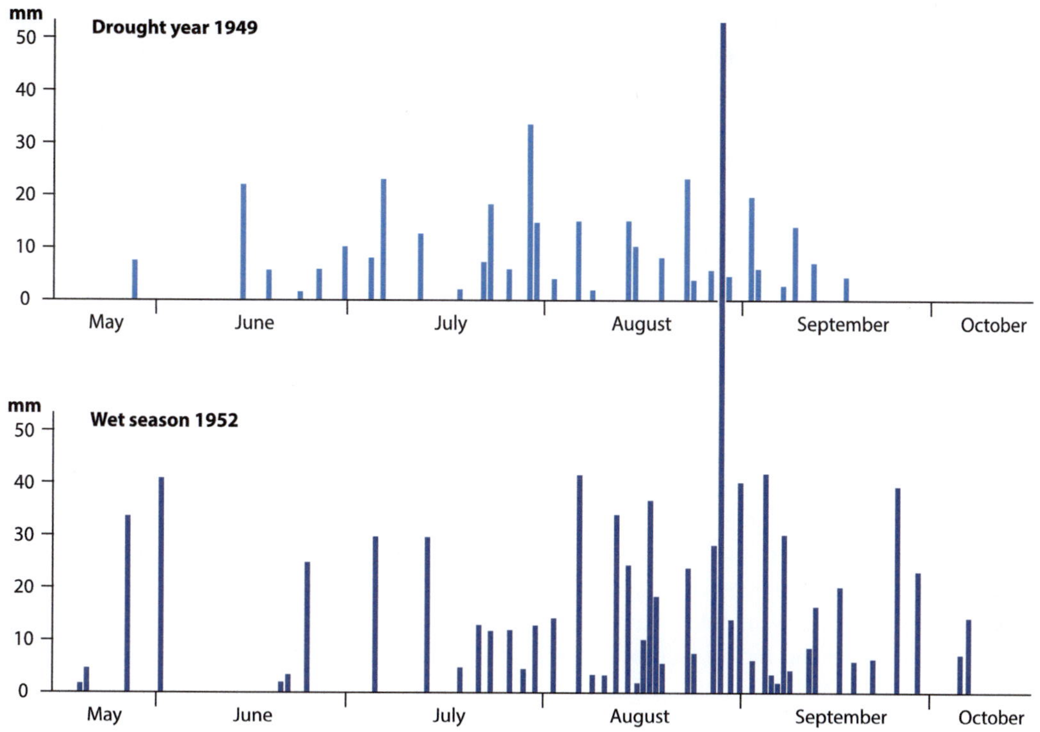


**Fig. 5.14** Daily precipitation in Niamey, 1949 and 1952. Comparison of an above-average wet year and a dry year. *Source* Janke (1973)

in the southern Central Sahara of Northeast Niger, which, however, did not bring any measurable precipitation, but only individual raindrop falls (see Grunert 1979). Rain collectors, which receive precipitation from both regimes, both from the Mediterranean winter rains and from the tropical summer rains, are the Central Saharan mountains Hoggar and Tibesti. With a little more than 100 mm of precipitation in the long-term mean, they are moisture islands in the desert.

Unusually high is the potential evaporation with values of 2,500–3,000 mm per year. The evaporation rate in the Central Sahara shows a clear central-peripheral gradient. Thus, in Colomb Béchar (Algeria), which is about 430 km from the Mediterranean coast, 4,000 mm are measured, in the also Algerian Adrar, about 850 km south of the coast, already just over 5,000 mm. In the even drier eastern Sahara, values of up to 6,000 mm are given. The mean relative humidity in winter is 40–55 per cent and only 20–25 per cent in summer. In the southern central Sahara, values below 10 per cent were measured several times in April during our research work from 1981 to 1990.

Also, the eastern part of Libya's Cyrenaica and all of Egypt belong to the Saharan dry climate. The precipitation is only slightly higher than in the interior. The winters are mild, while the summers, from May to September, are characterized by hot and dry conditions. Rainfall amounts are low and more variable. Not so extreme is the climate of the Nile Delta region. In Alexandria, 191 mm of precipitation falls from November to February; in Cairo, it is still 24 mm, with average temperatures of 13.3 °C in January and 28.3 °C in July. The precipitation in the Delta comes from northeastern wind-flows, which alternate with hot, dry, and often dusty southwestern winds originating from the Sahara. Since these blow about 50 days a year, on two to three consecutive days, they are referred to in Egypt as Khamsin (or Chamsin="50"). The Red Sea coast in the region of Port Sudan gets somewhat more rain. There, from October to February, northeastern winds force the moist air masses to rise at the escarpments of the Red Sea trench.

Between southern Morocco and Mauritania, the coastal desert of Western Sahara seamlessly adjoins the Sahara to the west. This is related to the cold Canary Current offshore. The precipitation is thus similarly low as in the central Sahara. Nouadhibou in Mauritania receives a mean of 27 mm of annual precipitation on five rainy days a year.

East Africa occupies a climatic special position compared to the other intertropical regions of West and Central Africa. The altitude of the highlands of Tanzania, Uganda, Kenya, and Ethiopia and the morphology of the volcanic cones and rifts, as well as the large water surfaces of the lakes and the broad coastal slope of Somalia on the Indian Ocean, overlay the effects of the autochthonous precipitation regime of the zenith rains and cause non-zonal and small-scale circulation patterns that manifest themselves as erratic events. Thus, the annual precipitation in northern Kenya is less than 300 mm per year, although East Africa, due to its planetary location in its southern part, belongs to the wet equatorial tropics and its northern part to the wet-dry tropical climate. "The most impressive climatic anomaly in all of Africa" (Trewartha 1981, p. 134) will be explained in detail below using the four precipitation zones of East Africa.

The precipitation of 500 mm to more than 2,000 mm in the **East African highlands** depends on the positions of the sun and falls as zenith rain during the twice passage of the ITCZ. The first maximum from March to May occurs with the shift of the ITCZ to the northern hemisphere. The southeast trade winds then hit the coast at right angles with warm air masses and flow onto the highlands. In the northern summer, due to the low Indian monsoon, there is a change in the flow of the trade. They prevent the transport of moisture to East Africa and thus lead to a dry season from June to September. The only notable rainfall during this period occurs east of Lake Victoria, with its effect stretching as far as central Kenya. These regionally induced "continental rains" (Berger 1989) from the lake and the Congo Basin led to a three-peaked precipitation curve, for example, in Nakuru, Kenya. In September and October, after the end of the Indian Monsoon, East Africa comes under the influence of the northeast trades. If the trades blow stronger from north-northeast, i.e., from the Ethiopian Highlands, then they are dry, if they blow from east-northeast from the Indian Ocean, they are more humid and provide a small rainy season, followed by the dry months from December to March under the influence of the northeast trades. However, in Tanzania, south of the Equator, the southern summer rainy season begins in December. Although the highlands of Kenya receive annual precipitation totals of more than 760 mm across the board and thus clearly above the critical limit of the growing period, the variability of precipitation in space and time is exceedingly high here, too, and complicates the correct choice of seeding time. In the event of unexpected drought periods, the seed withers, and if the seed is too late, it may not mature (see Berger 1989). The coastal stations, on the other hand, receive just as much rainfall as the fringe of the Rift Valley in the west. In the northwest of East Africa, already in the Sudan zone, but also towards the Highlands of Ethiopia, the rainfall drops to values below 500 mm, for example, in Lodwar in northwest Kenya, it is down to 165 mm.

The very differentiated rainfall regimes of the Highlands of Ethiopia divide this, according to Hurni (1982), into three hygric regions. The rainiest regions are in the windward of the southwest monsoon, i.e., in western Ethiopia and western northern Ethiopia (for example, Addis Ababa), with a rainy season in the summer from March to November, which provides an annual mean of 1,500–2,000 mm of rainfall. Regions with two rainy seasons at the zenith passage of the sun in spring and autumn are found in southern Ethiopia. The amount of rainfall there significantly decreases from the west (high plateau of Gamogofa with 1,189 mm) to the east (Ogaden with only 186 mm). In central and eastern Ethiopia and the northeastern highlands, two rainy seasons occur, separated by a dry season. The southwest monsoonal summer rains from June/July to August/September provide 50–70 per cent of the total annual rainfall; the spring rains, which are caused by southern and eastern winds from the Indian Ocean, account for only 15–30 per cent.

The Somali Peninsula in the Horn of Africa receives 100–500 mm annual rainfall, which falls in less than 50 days as equinoctial rain during the passage of the ITCZ to the transitional seasons. Barely 100 mm of rain per year in less than ten days is received by the coastal area of the Gulf of Aden, which lies in the lee of the monsoons in summer and in winter is exposed to the dry, continental northeast trades. The low-lying coast region of the Red Sea represents the fourth rainfall zone of East Africa. In the north, the Sudanese-Eritrean coastal, including, Port Sudan and Massaua, experiences a winter rainfall regime, whose low rainfall results from the trade winds that build up against the coast. In the hinterland of the Eritrean coast, on the northeast-exposed steep escarpment of the Ethiopian highlands, the overlapping of cyclonic winter rains and summer solstitial rains creates a regional favor space.

The mountainous regions and mountain massifs of High Africa belong to the **tropical Highland Climates** and show a vertical climate zonation. As in all tropical mountainous areas, thermal climate zonation can also be differentiated in East Africa due to the decrease in temperature with elevation. In the humid-temperate altitude zone, which lies between 1,100–1,200 m and 2,200–2,400 m a.s.l. and corresponds to the "Tierra templada" (altitude zone of the tropical Andes), the annual mean temperature drops to 17–18 °C up to 2,000 m altitude, with mean daily maximums of about 20 °C and minimums of only about 11 °C. The temperature curve prevents the transmission of tropical diseases such as malaria or sleeping sickness and allows cattle farming due to the absence of the tsetse fly (*Glossina morsitans* does not occur here). With simultaneous frost-free conditions, agricultural use with coffee and tea plantations is just as possible as the cultivation of non-tropical field crops, vegetables, and fruit. In the Highlands of Ethiopia, between the "hot land", **Bercha** or **Lower Kolla**, up to approx. 1,500 m. with annual mean temperatures around 30 °C and more, the upper or actual Kolla between 1,500 and 1,800 m with a mean annual temperature of about 26 °C, and the "wine country," the **Weyna Dega**, between 1,800 and 2,500 m with a mean annual temperature of around 22 °C (Fig. 5.15). While cotton and sugarcane are grown in the Kolla, the Weyna Dega is the cultivation area for tropical highland cultures, such as coffee, teff, citrus fruits, and corn. The most productive growing areas are, therefore, not in the hot lowlands, but rather in the cooler middle altitudes (Addis Ababa at 2,450 m). This applies to all East African highlands (Nairobi 1,798 m, Lake Victoria 1,134 m).

At the upper limit of the "African Tierra templada", at 2,200–2,400 m altitude, the highest values in East Africa are reached with mean annual rainfall totals of 2,200–2,400 mm. Especially on the windward sides of the volcanic massifs, orographic rainfall leads to extremely high values. Thus, the rainfall on the west side of the Ruwenzori massif reaches around 4,000 mm, and on the southeast side of Kilimanjaro, more than 2,500 mm.

Between 2,200–2,400 m and 3,700–3,800 m altitude extends the "Tierra fria" ("cold land") of East Africa, with over ten humid months and a mean annual temperature of 6–16 °C. In Ethiopia, this altitudinal zone ranges from 2,500 m to about the tree line (**Upper Dega**, 3,600–3,800 m) and is the cultivation zone for non-tropical field crops such as barley, oats, and wheat. The volcanoes superimposed on the highlands finally reach into the **Werch zone** above 3,500 m, where, at annual temperatures <14 °C, only pasture use is possible. The daily temperature range is smaller compared to the lower altitudinal zone. These altitudinal zones are important because they are not only found on the outer escarpments of the Ethiopian Highland but due to the fragmentation of the highlands into individual high plateaus ("Ambas"), they can be found everywhere in the highlands. The total rainfall decreases with increasing altitude up to about 3,000 m. The

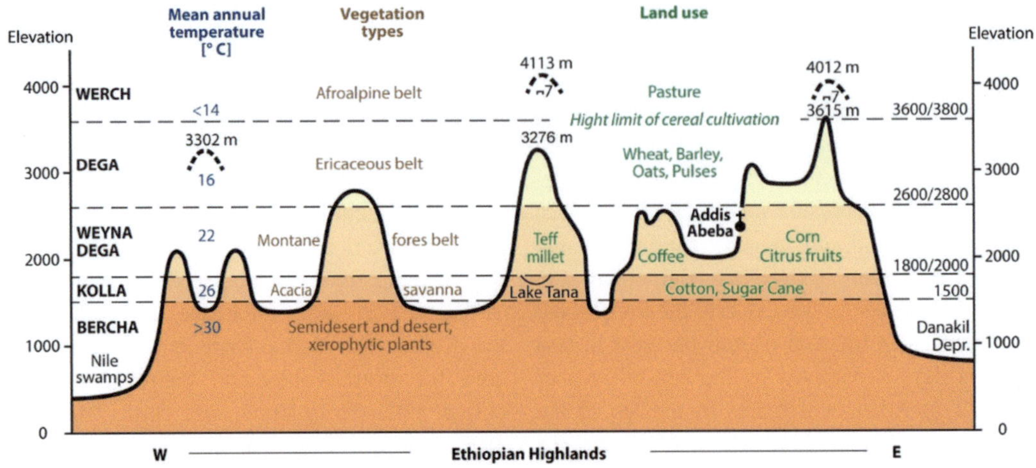


**Fig. 5.15** Altitudinal zonation in the highlands of Ethiopia. Modified from Hurni (1982) and Weischet and Endlicher (2000)

maximum rainfall, exceeding 1,500 mm annually, occurs at an elevation of 3,600 m. Above 4,000 m, snow falls periodically, episodically at lower altitudes. Frost occurs in the Werch zone frequently throughout the year, predominantly in the dry season. In the alpine mountain areas of East Africa, such as the Ruwenzori massif, the Uvira plateau of the Central African Ridge in the east of the Democratic Republic of Congo, and the Virunga volcanoes, the "Tierra helada" reaches up to 3,800 m. The characteristic Afro-Alpine Paramo vegetation characterizes this altitudinal zone with mean temperatures of the warmest month below 10 °C and a climatic snow line on average of about 4,300 m. At the Ruwenzori massif, it is at 4,600 m.

In **southern Africa**, there are **five climatic zones** (regions or provinces). From the Equator to 20° S, the **wet-dry tropical climate** extends with the characteristic zenithal rainfall. South of 20° S, a **dry subtropical climate**, dominates on the west side of the continent, followed to the east on the plateaus of the central highlands by a **dry subtropical climate** with precipitation in summer that transitions to the east and southeast into a **wet subtropical climate**. The southern tip of the continent just reaches the **Mediterranean subtropical climate**. The orography and the two coastal ocean currents, the Agulhas and the Benguela Currents, significantly influence the climatic conditions of southern Africa. Southern Africa, as part of High Africa with an elevation from 1,000 to 2,000 m, is enclosed by the approximately 5,000 km long Great Escarpment. The Great Escarpment is often preceded by narrow coastal strips and only isolated wide coastal plains and mountainous coastal regions. Central inland basins form the sand and Bushveld of the Kalahari at 900–1,200 m altitude, and the Okavango Basin to the north that follows. In the east, the Highveld plateaus of southern Zimbabwe and Transvaal (today Limpopo Province) rise to an altitude of 1,200–1,800 m, in the south, the Upper Karoo, and in the west, the southwestern African peneplains. The coastal plain is particularly widespread in southern Mozambique, while the coastal terraced landscape of KwaZulu-Natal and Eastern Cape (formerly Ciskei and Transkei) are only narrowly developed. At the southern tip of the continent, the Great Escarpment retreats about 200 km inland. To the south, the Cape Fold Belt (Swartberge up to 2,324 m) and the Great Karoo follow. In the southwest, the coastal area between the Atlantic and the escarpment lies the Namib Desert, a coastal fog desert, which is partly caused by the cold upwelling water of the northward flowing Benguela Current. The average water temperatures are 14 °C in July and 18 °C in January. On the other hand, on the east coast, the water temperatures are quite high due to the southward-directed Agulhas Current, with about 25–26 °C in January and 21–22 °C in July. It, therefore, acts practically year-round as a "heating" for the coast.

The **wet-dry tropics** form the transition area between the Congo Basin in the north, the center of the dry zone in the Kalahari in the south, and the Namibian coastal desert in the west. They can be divided into three climatic regions (subzones). The southern border area of the Congo Basin, the Angolan highlands, the Lunda Ridge, Zambia, as well as the northernmost parts of Zimbabwe and Namibia, are subject in the summer half-year to the humid Atlantic-Equatorial Congo air masses with a precipitation gradient from 1,500 mm in the north at the fringe of the Congo Basin to under 500 mm at the Tropic of Capricorn in the south. The rainy season in Zambia follows a regional pattern. In the north (10° S), it begins at the end of October and lasts until the end of April. However, the south of Zambia (18° S) experiences a shorter rainy season, starting in mid-November and ending in mid-March. The mean daily maximums exceed 30 °C in Lusaka in the dry season and still 25 °C in the rainy season. Of great importance for the moisture transport in this part of the continent is, next to the ITCZ, the second most important convergence zone over Africa, the confluence over East- and South Africa between the monsoon of the Atlantic Ocean and trades of the Indian Ocean. It is also known as the Congo Air Mass Boundary ("Congo Air Boundary"). It is located in the southern summer over the Angolan highlands. There, a thermal low forms, which extends into northern Namibia (Ovamboland). In the transitional seasons and the northern summer, it often runs across the continent to the East African and Ethiopian Highlands (see Fig. 5.16).

The dry subtropical climate with precipitation in summer occurs in southern Zimbabwe, Botswana with the Kalahari and the Okavango Basin, as well as the highlands of central Namibia. The precipitation yield decreases towards the south with the slowly diminishing influence of the equatorial air and the increasing dominance of the subtropical high-pressure belt. Due to the onshore southeast trades, whose dryness is further enhanced when crossing the continent by the foehn effect of the eastern escarpment, the frequency of precipitation also decreases towards the interior of the continent from east to west. The annual precipitation in Mutare in Zimbabwe (20° S) is 756 mm, in Bulawayo, 589 mm, in Maun, Botswana, 471 mm; and in Tsumeb, Namibia, 553 mm. The highlands of the west and east are already considered high-altitude climates. Harare (1,470 m) and Windhoek (1,676 m), therefore, have an annual amplitude of at least 12 °C and are hardly heat-stressed. The coastal lowlands of central and southern Mozambique receive just like the coastal lowlands of Madagascar due to the year-round dominance of the lapse of the onshore southeast trade winds annual precipitation of more than 1,000 mm, which is also influenced by the southern summer shift of the ITCZ. High temperatures with mean temperature maxima, such as in Mossurid (15° S on the Strait of Mozambique) of more than 30 °C, and high mean air humidity of over 80 per cent in the rainy season led to a permanent sultriness. Of particular interest are the climatic conditions in Madagascar. Due to the location of the mountains running across the stream of the trades, the island is divided into an extremely rainy eastern half with precipitation over 2,000 mm and a western half in the lee, whose rainfall, additionally reduced by frequent foehn, drops to 500 mm.

The **dry subtropical climate of the central plateau landscapes** of southern Africa is characterized by strong continentality and a pronounced climatic west–east contrast. Geographically, the escarpment highlands of southwest Africa (1,400–1,800 m) belong to this climatic region. As a dividing line between Kalahari Sandveld or Bushveld in the north and the Upper Karoo in the south at 800–1,200 m, the 400 mm isohyet is considered. It thus separates the dry western highlands climatically from the wetter eastern part. While in the east of the subcontinent, the Great Escarpment receives over 1,000 mm of annual precipitation, it is on

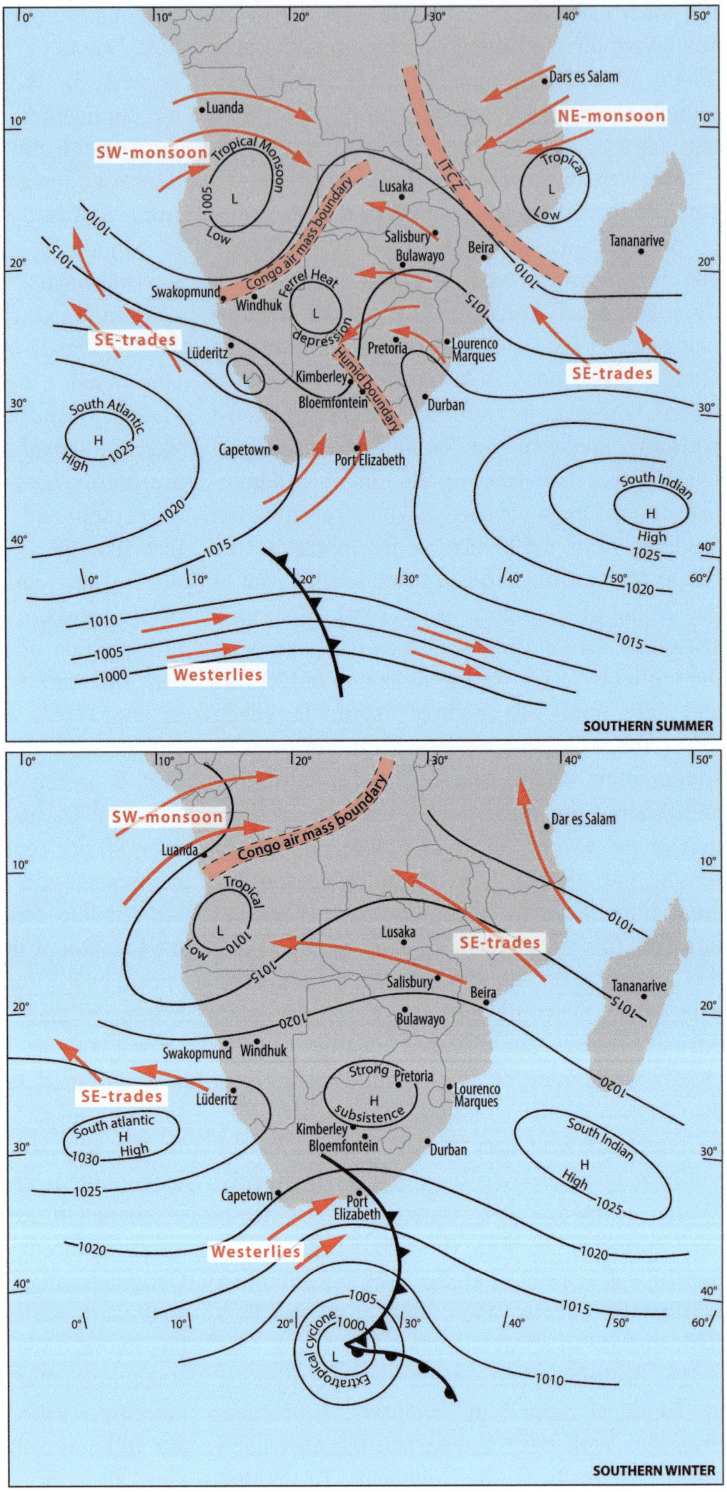


**Fig. 5.16** Atmospheric pressure and prevailing surface winds over southern Africa in summer and in winter. Pressure units are millibars reduced to sea level. *Source* Weischet and Endlicher (2000)

the eastern plateaus still about 700 mm, in the west only half (Windhoek 370 mm), and in the Namib less than 100 mm. The number of rainy days decreases from 100 to 120 in the eastern escarpment to 80 in the former Transvaal, 40 in the Kalahari, and under 10 in the Namib. In the north, the precipitation falls to over 80 per cent between October and March. On the East Coast, the rainfall activity regularly begins in August/September, while in the far west, erratic summer rainfall occurs for many years, often not until mid-January. For Namibian land users, the "farmer's rule" applies that a rainy season, in which the first rainfall does not occur until the end of January, is referred to as a "drought year" because the total amounts required for good pasture development can no longer be expected. Also, crops planted by farmers in North Namibia and South Angola are then largely considered ruined. At least four consecutive days of rainfall occur in Eswatini up to ten times a year, but in the northwestern Cape Province, not even once a year. Typical of the wet-dry tropical climate, the region experiences significant temporal and spatial variability of rainfall from year to year. It increases as mean rainfall decreases towards the west and repeatedly leads to extreme drought periods. In the central and western Karoo area alone, over 60 drought months were counted between 1926 and 1939. The detailed statistical analyses by Kempf (2000) based on 64 centennial measuring stations show that not only the number of drought months increase towards the west, but also the statistical mean rainfall becomes increasingly irrelevant. The long-term average rainfall is only reached in one out of ten years, whereas seven drought years are compensated by two extremely wet years. Recent studies show that the fluctuations are also linked to the ENSO phenomenon (see Excursus in Chap. 9). In the dry, long-lasting high-pressure current weather of the winter half-year, frost occurs down to −5 °C, at ground level also down to −10 °C, especially affected are the valleys in the border area of the Great Escarpment. In the southeastern half of the central highlands, the annual frost period lasts 120 days, in Lesotho, 180 days, and at elevations above 1,500 m, ground frost occasionally occurs even in high summer. Particularly problematic for land use are cold snaps of polar air masses, which lead to radiation frost, the so-called "black frost", and can cause considerable damage to corn and wheat plants.

West of the Great Escarpment dominates the **dry subtropical climate** of Namib. Like the South American Atacama, it is a coastal desert typical along the western sides of the continents, stretching 3,000 km from 9° to 33° S and located in the eastern part of the South Atlantic subtropical high-pressure belt. This high-pressure belt is particularly caused by the stably cold upwelling waters of the Benguela Current, which leads to a semi-permanent subsidence temperature inversion over the sea. This prevents any convection. The center of the Namib is in the southern tropic at Swakopmund, which receives only 1 mm of rainfall annually. Like all coastal deserts, the Namib is also a desert that experiences high frequencies of fog. It is usually covered by a dense fog blanket in the morning, which only dissolves in the midday hours and leads to a year-round decrease in the radiant intensity. For Swakopmund, this means mean monthly temperatures of 12.6 °C in August and 18.5 °C in February, with a mean interior diurnal temperature range of just under 6 °C. Extreme values of over 40 °C can occur in exceptional situations, for example, when a coastal low is forming due to easterly foehn-like winds. These violent mountain winds occur regularly in August and September and cause enormous dust transport from the dried-up rivers (Riviere) to the ocean.

In the extreme southeast of the African continent, south of the tropical coastal lowland of Mozambique, the **east side climate of the moist subtropical climate of the south and southeast coast** follows. A high humidity with maximum rainfall in summer occurs in the coastal strip from Maputo to Port Elizabeth. Annual precipitation at the coast reaches 750–1,250 mm and increases towards the escarpment up to 1,750 mm. The Great Escarpment itself reaches about 1,220 m in height at the Transvaal Escarpment and receives up to 3,200 mm of rainfall. It separates the coastal and lowland areas from the eastern Highveld not only morphologically but also forms a pronounced climatic divide in whose lee the amounts of precipitation rapidly decrease. Thus, Mokhtolong in Lesotho only receives 586 mm annually, some of which falls as snow. Responsible for the precipitation are tropical and extratropical atmospheric disturbances from the northeast and southeast. In winter, high-pressure weather prevails, but even in this season, with the dominance of humid tropical air from the Indian Ocean, which is responsible for a year-round debilitating humidity in the coastal region, for example, in Durban. In the terraced landscape fronting the Great Escarpment, a stationary temperature inversion is more persistent more frequently in the winter half-year. It is about 1,000 m lower than the escarpment and ensures that only shallowly developed humid maritime air masses cannot sink in the interior. At the condensation level, the so-called "mist belt" leads to a montane rain forest, whose species spectrum and phenology show parallels to the Afro-montane altitudinal zone of vegetation.

In the extreme southeastern coastal section, the Eastern Cape Province between Cape Agulhas and Port Elizabeth, and in the hinterland of Mosselbay-Knysna, annual mean temperatures of 18–20 °C (moist subtropics) are achieved by the superimposition of tropical summer with extratropical winter precipitation from the cyclonic westerlies all-year precipitation heights of over 1,000 mm with maxima in spring and autumn.

The different patterns of temperature, precipitation, and atmospheric circulation, and, therefore, the completely different regional climates also shape agricultural use: Citrus cultures and sugarcane plantations dominate the coastal sections, but also the terraced landscapes located east of the escarpment, in contrast to the eastern parts of the interior highlands, where grain cultivation predominates.

At the extreme southwestern tip, the African continent just reaches into the characteristic **Mediterranean climate**. It influences the mountains of the western Cape (Cedarberg chains, Swartberge) and the Tankwa-Karoo. They are influenced in winter by cyclonic storms, and polar frontal precipitation reaches the area, which brings precipitation of 600–1,000 mm in up to five humid months. Snow depths of over three meters can be reached, for example, at the well-known Theronsberg Pass near Ceres. The landscapes in the lee of the mountain ranges are low in precipitation. In the highest elevations, such as the Hexrivier mountains (up to 2,250 m) and the southern frame mountains of the fruit-growing basin of Ceres, the precipitation often falls as snow, which is particularly important for water supply. Frost poses a significant constraint for agricultural use, as do drought periods during the summer dry season when trades from the south and southeast dominate. Typical agricultural crops are wheat, which is grown in rainfed agriculture, fruit, and increasingly vineyards in recent times. The natural potential vegetation of the Cape is subtropical sclerophyllous shrubs, such as the widespread sclerophyllous (maquis) or Fynbos. It forms the extremely rich endemism floral zone of the Capensis, whose genetic relationships extend into the Afro-alpine regions of the East African alpine mountains.

A subtropical winter-humid climate also prevails in the Mediterranean-influenced regions of the Maghreb countries Tunisia, Algeria, Morocco, and in the Libyan coastal plains, except for the Syrte Arc. To the south and southeast, the Maghreb mountains of the Rif and the (Middle, High, and Anti-) Atlas form the southern border of the "Mediterranean climate" to the dry subtropical climate of the Sahara. In the summer half-year, the Maghreb states are under the influence of a subtropical high-pressure belt with radiation climate and dryness in up to six arid months. The winter precipitation comes from eastward drifting midlatitude low-pressure systems of the equatorward shifted westerlies and yields annual totals of 600 mm to a maximum of 2,000 mm of precipitation on the Atlantic western sides of Rif and Atlas in Morocco. Up to 1,800 mm are also reached on the northern slopes of the Mediterranean coastal mountains of Algeria. The mean annual totals regularly decrease from north to south up to the border of rain-fed farming at 300–400 mm. In the same ratio, the annual precipitation variability increases, comparable to the dry areas of southern Africa. Precipitation often occurs in the form of intense, heavy rainfall (at least 30 mm in 24 h). In Tunisia, 25–50 per cent of the annual precipitation falls as such torrential rain and often leads to catastrophic floods (see Excursus: The flood disaster in Tunisia in fall of 1969). However, rainfall is subject to great variability annually, even in the winter half-year. In the greater heights of the mountains, there is snowfall. Frost and, to an even greater extent, vulnerability to drought are significant risk factors for agricultural use. In Djelfa in the Sahara Atlas at an altitude of 1,153 m, up to 82 frost days per year were recorded.

Atlantic climatic influences are only effective directly on the Atlantic coast of Morocco. Thus, in the northern Moroccan Meseta, 400–600 mm of annual precipitation falls with a maximum of 150 dry days. In particular, the cool coastal Canary Current with maximum temperatures of 20 °C in summer contributes to an isothermal climate with a diurnal range of only 6–10 °C. The Canary Current causes high frequencies of dew and fog, which influences vegetation up to 200 km into the hinterland along the Atlantic coast from Morocco to Mauritania. As mentioned, this Atlantic coastal section genetically represents a coastal desert, which, however, is dominated by the hyperaridity of the Sahara.

The morphological boundary between the Mediterranean climate and the climate of the Sahara is formed by the Sahara Atlas. Characteristic of the Saharan influence on the Mediterranean area are the sandstorms, such as the hot, dry, and dust-laden Sirocco (Scirocco), which is referred to as Ghibli in Algeria, Tunisia, and Libya, and as Scherqi in Morocco, and can lead to a rise in temperatures up to 50 °C in summer.

**Excursus: The Flood Disaster in Tunisia in Autumn 1969**

The worst flood disaster in recent times in Tunisia occurred in 1969. The consequences were not only measurable in over 500 fatalities and around 100,000 homeless people. Many houses were destroyed, traffic and transport routes (roads, railway embankments, bridges, and Ford routes) were interrupted, and the mining industry, as well as the water and energy supply of further regions, were paralyzed. These damages exceeded the capabilities of the Tunisian state for rapid disaster aid, so numerous foreign rescue actions had to be initiated. For example, from Germany, seven army helicopters participated in an airlift for supply in October 1969. The population is trapped by masses of water in the interior of the country.

The flood areas were distributed from north to south across eastern Tunisia (Sahel) from the Medjerda Valley to the Gabes Oasis. While floods in flat valley areas (for example, the Medjerda) are not uncommon because they are part of the natural catchment areas, it became apparent in the fall of 1969 that wide marginal plains were overflowed by high flood waves. The result of this was that, for example, the city of Kairouan was completely isolated from the outside world from September 26–29, a situation that briefly recurred on October 6. The sebkhas, i.e., the drainage-less, sometimes dry inland lakes, as well as the terminal lakes (for example, Kelbia in the Kairouan Basin), significantly increased their area. Lake Kelbia created an outlet towards the Mediterranean with the destruction of all main roads and bridges in its path. This stream, which lasted for several months, carved a new riverbed all the way to

Hergla (north of Chatt Meriam, Port El Kantaoui). A similar event had already occurred in 1932 when the Kelbia Basin also drained into the sea.

Apart from the flood in the Medjerda valley, especially above the delta, as well as in the coastal areas in the hinterland of Bizerte, on the Gulf of Tunis, Hammamet, and Gabes, the high flood waves on the gently sloping foothills of the mountain ranges were a characteristic feature. Unfortunately, it is precisely in these regions where the centers of Tunisian grain farming are located. The flood disaster almost completely destroyed the entire sowing and covered it with sand and rubble.

The disaster was caused by the intrusion of large amounts of polar cold air masses from the west, which was able to advance to the Gulf of Gabes during the course of September 23 and 24. This cold air, however, met very dry and warm air masses that came from the Sahara and lay over the Syrte area in front of Libya and Tunisia. In the contact area of both air masses over the Syrte, an enormous moisture absorption took place, and massive amounts of moisture-saturated air could move towards the mainland, where they precipitated in abnormal rainfalls.

This stable weather situation persisted from September 28 until October 5. As a result, the normally short-term heavy rainfall extended over several days and led to high precipitation totals at the end of September and the beginning of October. Such east weather conditions, also called "Chergui" situations, are part of the typical weather events in Tunisia towards the end of summer but usually do not lead to such extraordinary precipitation. Due to the high moisture supply from the Syrt, there were several heavy rains and various flood waves both in central Tunisia and in the coastal area From September 25–27, on October 6 and 7, on October 22, on October 26 and 27, and finally on October 29 and 30, 1969, this event occurred in various forms, with the first phase in September being the largest and most dangerous. After a long summer of heat and drought, suddenly concentrated heavy rain with violent thunderstorms set in on September 24 and 25. It spread to the entire Tunisian Sahel region, the lowlands, and the central Tunisian plateau. Within three days, rainfall amounts of 200–300 mm, locally even more, fell. At almost all weather stations, half of the total annual rainfall was reached within these three days. Towards the south, some stations almost reached the complete annual average: El Djem measured 218 mm of precipitation in 24 h (annual average: 275 mm) and the Domaine Saint Louis station received a new absolute daily maximum of 225 mm in this region.

| Mean annual precipitation 1901–1960 | Kairouan 295 mm | El Djem 275 mm | Sousse 324 mm | Maktar 500 mm | Bizerte 642 mm | Tunis 441 mm |
|---|---|---|---|---|---|---|
| Mean precipitation September 1901–1960 | 38 | 34 | 49 | 41 | 34 | 33 |
| Precipitation September 1969 | 144.9 | 408.2 | 130.3 | 418.1 | 15.6 | 49.1 |
| Precipitation September 25th–27th 1969 | 78.7 | 319.0 | 71.8 | 315.5 | 5.6 | 0.5 |
| Mean precipitation October 1901–1969 | 35 | 37 | 45 | 41 | 70 | 55 |
| Precipitation October 1969 | 427.8 | 374.2 | 431.5 | 439.6 | 186.7 | 184.4 |
| Precipitation September and October 1969 | 572.7 | 782.4 | 561.8 | 857.7 | 202.3 | 233.5 |

While in September, the rainfall fell exclusively in "Chergui locations", a northwest wind also set in October so that the north also recorded heavier rainfall. For some representative stations, the September and October rains of 1969 are compared here with the long-term means: Capturing the actual flood masses that occurred proved extremely challenging. It can, therefore, only be rough calculations, not measurements, as most gauges were already destroyed at the beginning of the event. A comparison of the two most important (dangerous) flood bringers in the Kairouan Basin (Oued Zeroud and Oued Marguellil) with theoretically calculated maximum discharge donations and the water quantities that occurred from September 25–27, however, indicates the significance of this flood disaster: It shows that the September floods in Tunisia have far exceeded a "century flood" and the effects were catastrophic (Mensching et al. 1970).

On September 22, 2018, a Chergui (east) weather situation, which is part of the typical weather events in Tunisia towards the end of summer (see above), led to record rainfall with catastrophic consequences in the eastern area of Tunisia, in Nabeul and Sousse. In Nabeul, 55–297 mm of precipitation was measured within 24 h, in Sousse, there were still 48–155 mm. The long-term mean annual precipitation in Sousse is 339 mm, and in Nabeul 404 mm.

## 5.3 Climate of Africa: Favor and Risk

Africa's climate regions reveal a distinctly unique climate compared to other areas of the Earth's surface. Notably, Africa exhibits some deviations from the standard global pattern of climatic distribution. On a continent with a rapidly growing population, where over 70 per cent of people rely directly on agriculture for their livelihood, understanding the hygric and thermal conditions is essential for informed land-use decisions. On the other hand, the agro-climatic database in the majority of climate regions shows that it is not sufficient to simply use the mean annual precipitation totals or the annual mean temperatures for assessing the agricultural potential, the use of water resources or the implementation of infrastructure measures, in addition to the other physiogeographic features such as relief, soil, etc. Rather, in addition to the annual precipitation and temperature over Africa, the local, regional, and temporal rainfall variability and reliability must be considered, as well as the sometimes-large fluctuations that can occur from year to year.

The rainfall of the wet equatorial tropics on the Guinea coast and the Congo Basin, with more than 1,500 mm annually, is sufficient for the formation of the tropical rainforest. Despite remarkably differently developed rainfall regimes, the highlands of East Africa, including Kenya, Uganda, and Ethiopia, receive ample rainfall, making them suitable regions for arable farming and livestock production. This circumstance in Ethiopia, in particular, has led to the purchase or lease of huge land areas ("**land grabbing**") of up to 300,000 ha by agricultural investors from India and China, among others. These large farms promise millions in profits from the export of food, a not entirely unproblematic undertaking. In contrast to the equatorial latitudes In West and Central Africa, where a dry season with up to 34 days without any precipitation can occur from December to February, which, on the other hand, does not affect the local agricultural use, the variability of precipitation in East Africa is considerable, both in terms of the spatial distribution of rainfall and the certainty of the onset of the rainy season. The variability of rainfall makes it difficult to choose the right sowing date. In the event of unexpected drought, the seed withers; if sowing is too late, it does not ripen. Delays in the start and changes in the duration of the rainy season, as well as an unfavorable spatial and temporal distribution of rainfall within the rainy season, are increasing in the wet-dry tropical climate of the Sudan zone, the dry tropical climate of the Sahel in Africa north of the Equator, and the southern fringe of the Congo Basin, the Angolan highlands, the Lunda Ridge, Zambia, the northernmost parts of Zimbabwe and Namibia, but also southern Zimbabwe, Botswana with the Kalahari and the Okavango Basin, as well as the highlands of central Namibia in the southern hemisphere. The variability of annual rainfall increases as mean rainfall decreases, with decreasing amounts, and is in the range of the critical limit of the growing period at 30–40 per cent. Beyond the critical limit of the growing period, which is less than 500 mm average annual rainfall (with summer rain) in about four humid months, drought and aridity are decisive risk factors for agricultural use and thus affect around 40 per cent of the African continent. 500 mm is considered in the tropics as the minimum annual rainfall from which reliable **rainfed farming** can be operated.

This value is therefore referred to as the critical limit of the growing period (see Fig. 5.17 and, among others, Jätzold 1991). Through field cultivation techniques such as **dry farming** (in East and South Africa) or special varieties, for example, in millet in West Africa, rainfed farming can be operated in the regions with rainfall in summer up to annual mean rainfall of 350 mm, in the areas with rain in winter up to 250 mm (Jätzold 1984), however, always against the background of the risk of **rainfall variability** (Figs. 5.18 and 5.19).

The **Aridity Index (AI)** describes the ratio between average annual rainfall and potential evapotranspiration. The following applies:

- AI < 0.05 = hyper-arid; rainfall variability 100 per cent.
- AI 0.05–0.20 = arid; annual rainfall with winter rain up to 200 mm, with summer rain up to 300 mm, rainfall variability 50–100 per cent.
- AI 0.21–0.50 = semiarid; annual rainfall with winter rain up to 500 mm, with summer rain up to 800 mm, rainfall variability 25–50 per cent.
- AI > 0.50 = humid; transitioning to arid = dry-subhumid with AI from 0.50–0.65, rainfall variability <25 per cent.

The strong variability of rainfall at the critical limit of the growing period in the Sahel of West Africa in the twentieth century is evident in Fig. 5.20 using the example of Tillabéry in West Niger. Above-average rainfall in the 1930s, 1950s, and 1960s, as well as pronounced droughts in the 70s, 80s, and 90s of the twentieth century, are shown in Fig. 5.21.

The wet-dry tropical climate is subject to years of devasting drought and abnormally high rainfall that can result in severe floods. The

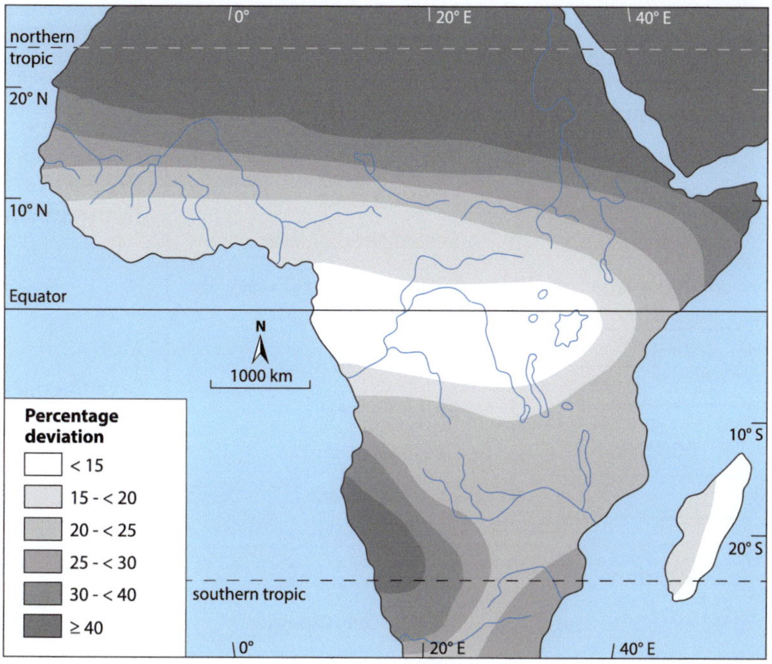


**Fig. 5.17** Change of mean annual precipitation rates, expressed as a percent change from the long-term mean

variability is a permanent feature of this climate. Plants and animals inhabiting this region have adjusted to the natural rainfall, with one exception: the human species (see Excursus Desertification). Droughts represent the largest (agro-)climatic risk factor in Africa. The complex relationships that lead to droughts in the Sahel have been intensively studied since the famines of 1968–1973 and the 1980s. In addition to natural autochthonous and allochthonous climatic aspects, anthropogenic factors such as overuse, deforestation, and overpopulation must also be considered, which intensify the effects. Since the UNCOD (United Nations Conference On Desertification) conference in Nairobi in 1977, which was convened by the UN after the drought disasters in the Sahel of Africa, these complex processes are referred to as Desertification.

**Excursus: Desertification**

Desertification, along with its far-reaching economic and social consequences, is a major problem in large parts of the world's drylands. These regions are inherently vulnerable to desertification due to climatic factors. Predictions of climate development suggest changes in the two climatic parameters most significant for desertification: temperature and precipitation. The "World Map of Desertification" (1977) published by United Nations Conference on Desertification is based on the definition arrived at the Nairobi Conference of the same year, describing desertification as the reduction or destruction of the biological potential of regions, eventually leading to desert-like

| | Precipitation | **Wet months** | Land use | ***Climate type*** Vegetation |
|---|---|---|---|---|
| | 200 mm | | Livestock farming + Small millet species | |
| | | **2** | Pearl millet + Livestock farming | |
| *Rainfall variability about 30 %* | 300 mm | | Peanut + Millet | ***Dry tropical climate*** Thorntree savanna |
| | | **3** | Sesam seeds + Millet | |
| | 400 mm | | Sorghum + Millet, Peanut | |
| | | **4** | Batate + Millet | |
| *Agronomic dry line (~ 500 mm), Rainfall variability about 20 %* | | | Corn + Cotton, Millet | ↑ Savanna parklands and low savanna woodlands ↓ |
| | | **5** | Cotton + Millet | |
| | 600 mm | | Cassava + Corn | |
| *Climatic dry limit* | | **6** | Dry rice + Beans | |
| | | | Arboriculture (Mango) | |
| | 800 mm | **7** | Cassava + Corn, Rice | |
| | ↓ | ↓ | Millet + Corn, Cassava | ***Wet-dry tropical climate*** Savanna woodlands |
| | | | Yams + Cassava | |
| | 1200 mm | **9** | Rice (irrigated) | |
| | 1300 mm | **10** | Cassava + Yams | |
| | ↓ | ↓ | Yams + Cassava | ***wet equatorial climate*** Tropical rainforest |
| | 1800 mm | **11** | Arboriculture | |
| | | | Rice (irrigated) | |

**Fig. 5.18** North–south profile of traditional cultivation crops across West Africa. Modified from Gießner (1988)

conditions. The desertification process is largely triggered and driven by humans. Inappropriate land use, i.e., overuse of fields or pastures, deforestation, overexploitation of groundwater reserves, and inadequate irrigation techniques leading to salinisation and/or desiccation, will destroy the plant cover, degrade the soils, and reduce the availability and quality of water. Inhabitants of such regions face severe consequences, including loss of livelihood, heightened risk of famine, increasing poverty, and an increasing potential for crises and conflicts. It is a fact that desertification processes are a set of highly complex cause-and-effect interactions, different in each case and each region with respect to the factors and mechanisms at work, and thus not accessible by monocausal explanations.

The UNCCD states in a 2013 report that a total of 11.8 million square kilometers of areas classified as dry areas (excluding naturally hyper-arid areas) are affected by desertification processes. The affected areas include (a total of 70 per cent of all drylands). In Africa, the regions on the northern and southern edges of the Sahara, in East and Northeast Africa, particularly Kenya, Ethiopia, and the Somali Peninsula, as well as in southern Africa, the peripheral areas of northern Karoo, Kalahari, and Namib are most affected or at least severely threatened. These landscapes cover an area of about 5.3 million square kilometers, accounting for about

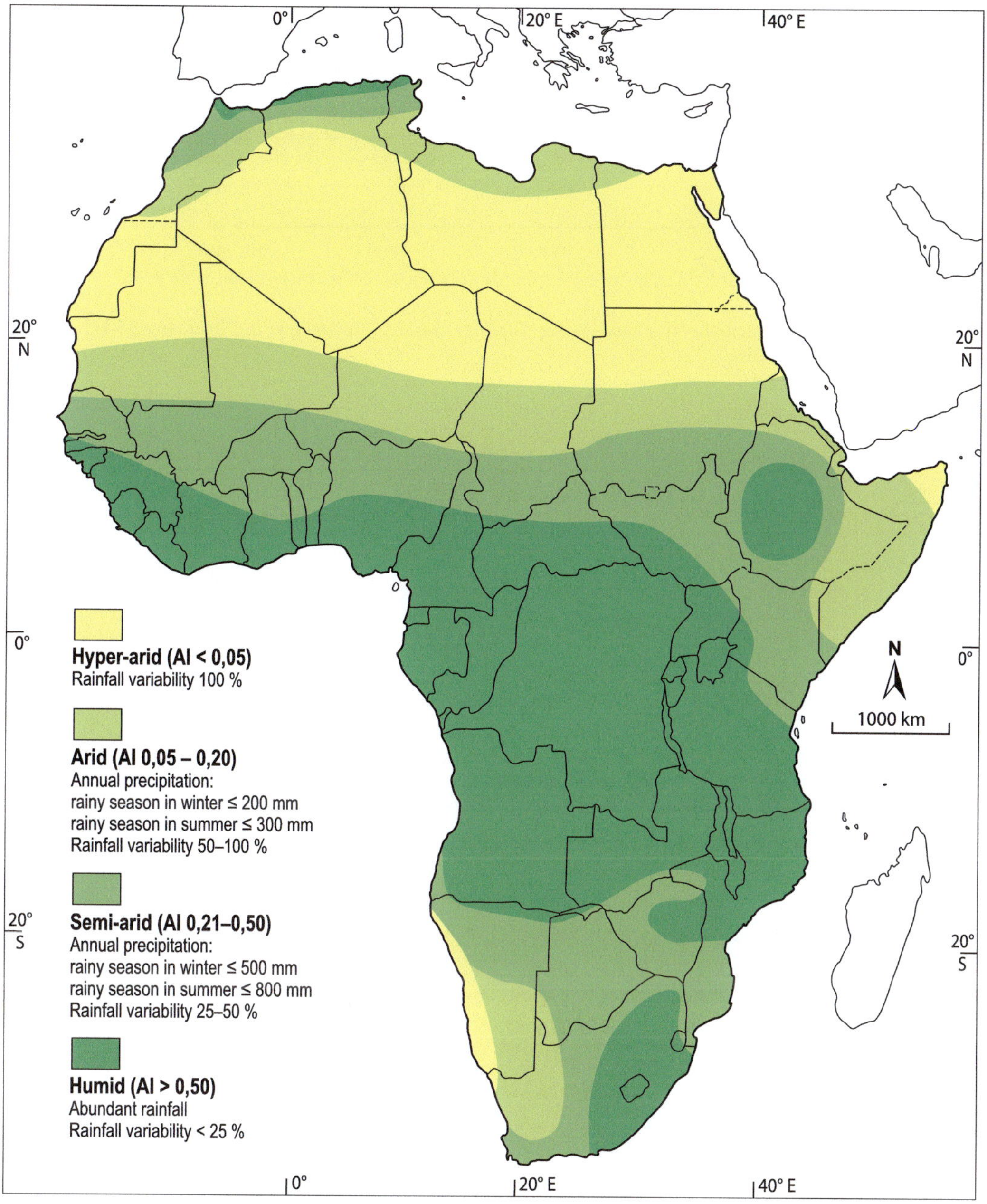


**Fig. 5.19** Aridity index in Africa. Modified from Wiese (1997)

30 per cent of Africa's area outside the natural deserts, with about 65 per cent of Africa's total population. In Central Asia, over 50 per cent of the land area of Kazakhstan, Kyrgyzstan, Uzbekistan, Tajikistan, and Turkmenistan are acutely threatened by desertification and landscape degradation. Also, in Latin America and the Caribbean, according to UNDP (2009), nearly 30 per cent of the dryland area (about 1.2 million square kilometers), which makes up 26.5 per cent of the total

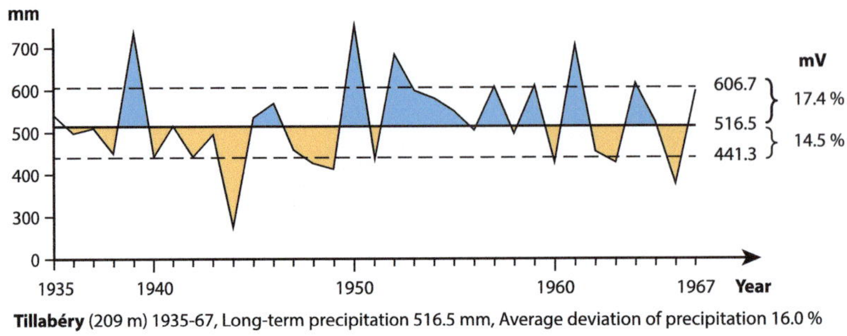


**Fig. 5.20** Annual Sahel precipitation anomalies for Tillabéry, western Republique of Niger, from 1935 to 1967. *Source* Mensching et al. (1969)

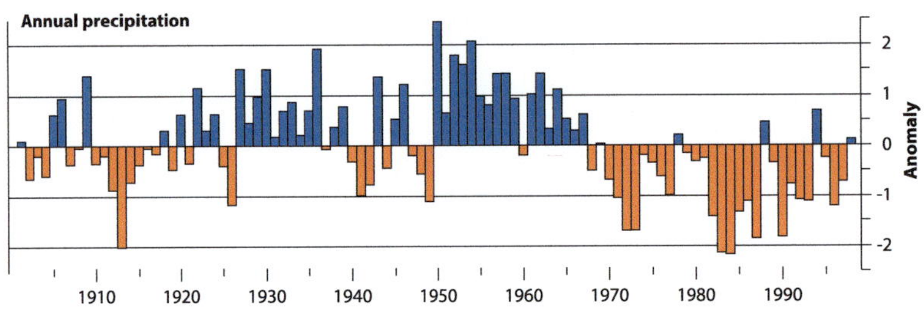


**Fig. 5.21** Mean precipitation anomalies between 1900 and 2000. Precipitation variations for stations in the Sahel (10–20° N) from 25° W to 30° E, expressed as an anomaly departure from the long-term mean. *Source* Brooks (2003)

area of Latin America and the Caribbean, is suffering from desertification processes. Desertification is also a problem in industrial and emerging countries. The affected countries include China, Argentina, Brazil, and Mexico, as well as the USA and some Mediterranean countries. Worldwide, over 2.5 billion people in 110 countries are affected or threatened by desertification. The dynamics of desertification are unbroken in large parts of the world and create alarming perspectives: If this trend continues, a decrease in agricultural land by two-thirds in Africa, by one-third in Asia (excluding China), and by two-fifths in Latin America is expected by 2025 (each compared to the reference year 1990). The fight against desertification is gaining importance, particularly due to the close links to global climate change, the loss of biodiversity, including genetic resources, and the resulting increasing food insecurity. Only a gentle use of resources and sustainable development in the drylands can contribute to combating desertification. For the first time, coordinated global action against desertification was established in the United Nations Convention to Combat Desertification in 1994 (see Chap. 7 Desertification).

The desertification circle of hell is probably unstoppable, and thus the natural climatic risk of drought, even against the backdrop of climate change, will not only intensify but probably multiply (see Chap. 9). In addition, increasing drought triggers local and regional self-reinforcing mechanisms, which particularly increase the pressure on the remaining vegetation. If this is further destroyed by overgrazing, it not only reduces evaporation but also increases

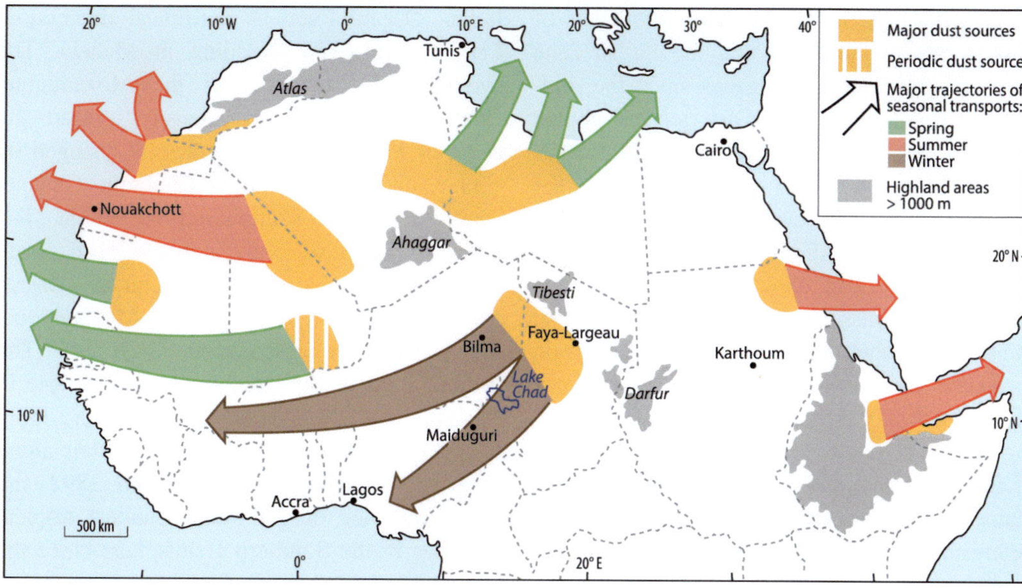


**Fig. 5.22** Model of Saharan dust sources with directions of seasonal long-range transport. *Source* Middleton (1986)

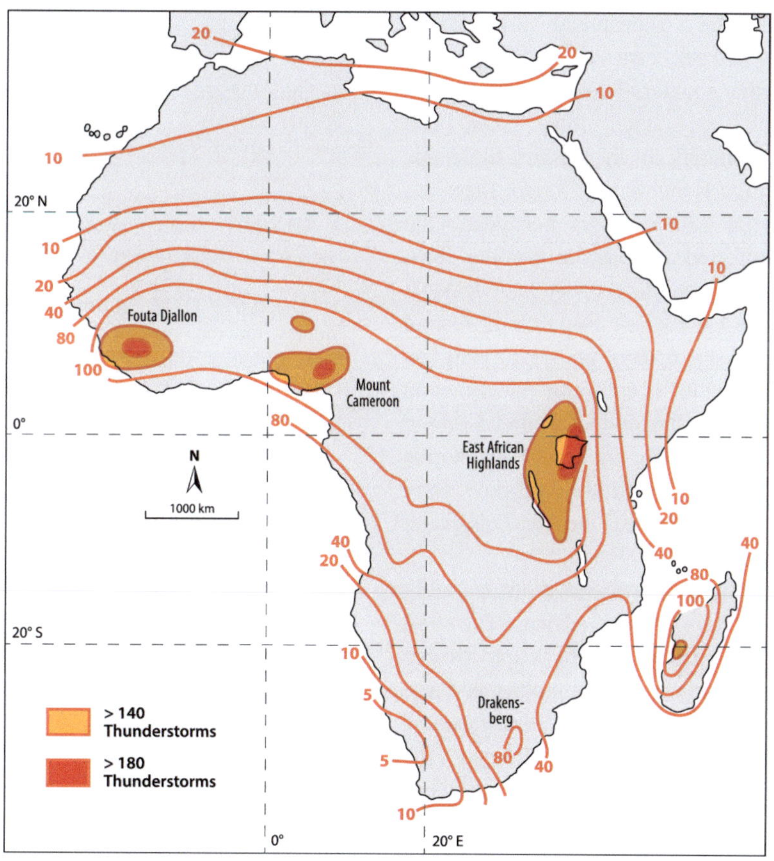


**Fig. 5.23** Mean annual frequency of thunderstorms in Africa. Modified from Griffiths (1972) in Goudie (1996)

the albedo and dust load. The frequency of dust storms in the Sahel and the Sudan zone has increased four to six times since the mid-1960s (Middleton 1986). Dust storms in and from the dry areas, especially the Sahara, as the largest global source region for dust in the atmosphere (possibly even up to ~50 per cent), represent another difficult to quantify climate risk. Assuming that about 60 per cent of the water vapor content for the summer monsoonal rains in the Sahel comes from the tropical rain forest and the savanna regions that lie between the west coast of Africa and the Sahel, and noting that in recent decades more than 80 per cent of the rain forest and savanna from Nigeria to Guinea has been destroyed, then the fear, that if deforestation continues, the monsoon circulation could collapse, is not unrealistic (Cook 1999; Fig. 5.22).

Other climatic risks include thunderstorms and point-by-point agriculture hail events. Thunderstorms are widespread in Africa. Approximately 20 per cent of the continent, especially in the equatorial zone of West and Central Africa, experience more than 100 thunderstorms annually. In the eastern highlands of the Democratic Republic of Congo, there are thunderstorms on 180–200 days per year. One of the most affected regions, if not the most thunderstorm-rich on earth, is likely to be Bukavu in the east of the Democratic Republic of Congo with 221 thunderstorm days per year. Hail, on the other hand, is not a common phenomenon in Africa, except for the region around Kericho in western Kenya. There, up to 80 hailstorms are recorded per year. Up to 100 hail days were counted on the high plateaus in KwaZulu-Natal and Lesotho (Goudie 1996; Fig. 5.23).

Tropical cyclones (hurricanes) occur rarely on Africa's coasts or over Africa. The east coast of southern Africa is hit by six to twelve storms per year. Only a small number of these storms reach up to 30° S and penetrate the inland. However, when such storms do cross the coast, copious rainfall and flooding occur. For example, one of these cyclones was Idai, the eleventh tropical storm and seventh cyclone of the 2018–2019 cyclone season in the southwest Indian Ocean. Even as a tropical depression, Idai caused severe flooding in Malawi. The cyclone then moved over the Mozambique Channel, increasing in intensity, grazed the coast of Madagascar, subsequently changed its course, and crossed the Mozambique Channel again, now in a southwestern direction. The intensity continued to increase with wind speeds up to 195 km/h. When it hit the African mainland in the region of Beira in Mozambique, Idai was the strongest cyclone since 2008. The cyclone's impact affected millions of people in Mozambique and Zimbabwe. There were more than 1,000 deaths. Thus, Cyclone Idai, along with two unnamed cyclones from 1892 and 1973, was one of the three deadliest tropical cyclones in the Southern Hemisphere since the beginning of reliable weather observations.

## References

Berger P (1989) Rainfall and agroclimatology of the Laikipia Plateau, Kenya. Geogr Bernensina, Africa Studies Series A 7

Brooks N (2003) Vulnerability, risk and adaptation: a conceptual framework. Tyndall Centre for Climate Change Research, Norwich (Working Paper 38), p 18

Cook KH (1999) Generation of the African easterly jet and its role in determining West African precipitation. J Clim 12:1165–1184

Flohn H (1965) ITCZ and meteorological Equator. Studies on the meteorology of Tropical Africa. Bonner Meteor. Abh. 5:49–57

Gießner K (1988) Naturressourcen und Unterentwicklung: Ökologische Aspekte als Rahmenbedingungen vor dem Hintergrund der Dependenztheorie. In: Hanns-Seidel-Stiftung (ed) Armut im Süden durch Wohlstand im Norden, Vilsbiburg, pp 143–166

Goudie AS (1996) Climate: past and present. In: Adams WM, Goudie AS, Orme AR (eds) The physical geography of Africa. Oxford University Press, New York, pp 35–59

Griffiths JF (ed) (1972) Climates of Africa. World survey of climatology, vol 10

Grunert J (1979) Wetter und Witterung in der zentralen Sahara 1977 und 1978. Würzb Geogr Arbeiten 49:1–65

Herbaud J (1966) Étude de ruissellement en zone urbaine à Niamey. Les bassins versants du Gounti-Yena. Bd. I & II, ORSTOM, Paris

Hurni H (1982) Hochgebirge von Semi-Äthiopien. Klima und Dynamik der Höhenstufung von der letzten

Kaltzeit bis zur Gegenwart. Geogr Bernensina, Africa Studies Series G 13
Indeje M, Semazzi FHM, Ogallo LJ (2000) ENSO signals in East African rainfall seasons. Int J Climatol 20:19–46
Janke B (1973) Naturpotential und Landnutzung im Nigertal bei Niamey, Rep. Niger. Jahrb. Geogr. Gesell. Hannover 1972
Jätzold R (1984) Das System der agro-ökologischen Zonen der Tropen als angewandte Klimageographie mit einem Beispiel aus Kenia. 44. Deutscher Geographentag Münster 1983:83–85
Jätzold R (1991) Eine Möglichkeit der angewandten Klimagliederung der Tropen und Subtropen jenseits der Regenfeldbaugrenze. Das Konzept der pastro-ökologischen Zonen. In: Endlicher W, Goßmann H (eds) Festschrift für W. Weischet. Freib Geogr Hefte 32:131–139
Kempf, J (2000) Klimageomorphologische Studien in Zentral-Namibia: Ein Beitrag zur Morpho-,Pedo- und Ökogenese. Diss. Uni.Würzburg, p 717
Mensching H, Gießner K, Stuckmann G (1970) Berichte. Geographie und Landeskunde 58(2):81–89
Mensching H, Gießner K, Stuckmann G (1969) Sudan-Sahel-Sahara. Jahrbuch der Geographischen Gesellschaft zu Hannover. Jahrbuch für 1969. Eigenverlag, Hannover, p 219
Middleton NJ (1986) Effects of drought on dust production in the Sahel. Nature 316:431–434
Trewartha GT (1981) The Earth's problem climates. University of Wisconsin Press, Madison, p 371
Troll C, Paffen KH (1964) Karte der Jahreszeitenklimate der Erde. Erdkunde 18:5–28
UNCOD (1977b) World map of Desertification 1: 25 Mill. Mit Explanatory Note, A/Conf. 74/2, Nairobi
UNDP (2009) Adapting to climate change through sustainable land management. Presentation at the UNCCD "Land Day" on June 6, 2009, a side event of the UNFCCC's "Bonn Climate Change Talks" (June 01–12, 2009). http://www.unccd.int/publicinfo/landday/docs/1TengbergUNDP.pdf
Weischet W, Endlicher W (2000) Regionale Klimatologie, Teil 2, die Alte Welt. Teubner, Stuttgart, p 626
Wiese B (1997) Afrika: Ressourcen, Wirtschaft, Entwicklung. Teubner, Stuttgart, p 262

## Further Readings

Collier P, Conway G, Venables T (2008) Climate change and Africa. Oxf Rev Econ Policy 24(2):337–353
Endlicher W (2000a) Afrika südlich des Maghreb. In: Weischet W, Endlicher W (eds) Regionale Klimatologie 2. Die Alte Welt. Teubner, Stuttgart Leipzig, pp 245–336
Endlicher W (2000b) Mittelmeergebiet. In: Weischet W, Endlicher W (eds) Regionale Klimatologie 2. Die Alte Welt. Teubner, Stuttgart Leipzig, pp 153–244
EUMETSAT (2014 ff) Europe's weather satellites delivering data in real time to the national weather services in the Member States. https://www.eumetsat.int/de
Giannini A, Biasutti M, Verstraete MM (2008) A climate model-based review of drought in the Sahel: Desertification, the re-greening and climate change. Glob Planet Chang 64:119–128
Hellmuth ME, Moorhead A, Thomson MC, Williams J (eds) (2007) Climate risk management in Africa: learning from practice. International Research Institute for Climate and Society (IRI), Columbia University, New York
Hulme M (1992) Rainfall changes in Africa: 1931–1960 to 1961–1990. Int J Climatol 12:685–699
Hulme M (1996) Climate change within the period of meteorological records. In: Adams WM, Goudie AS, Orme AR (eds) The physical geography of Africa. Oxford University Press, New York, 88–101
Hulme M, Doherty R, Ngara T, New M, Lister D (2001) African climate change: 1999–2100. Clim Res 17: 145–168
Jury MR, Enfield DB, Mélice J-L (2002) Tropical monsoons around Africa: Stability of El Niño-Southern Oscilllation associations and links with continental climate. J Geophys Res 107(C 10): 15-1–15-17
Jury MR, Rautenbach H, Tadross M, Philipp A (2007) Evaluating spatial scales of climate variability in sub-Saharan Africa. Theoret Appl Climatol 88:169–177
Knox J, Hess T, Daccache A, Wheeler T (2012) Climate change impacts on crop productivity in Africa and South Asia. Environ Res Lett 7:034032
Mason SJ, Jury MR (1997) Climatic variability and change over southern Africa: a reflection on underlying processes. Prog Phys Geogr 21(1):23–50
Mensching HG, Seuffert O (2001) (Landschafts-) Degradation—Desertifikation. Ein globales Umweltsyndrom. Petermanns Geogr Mitt 145(4):6–15
Nicholson SE, Selato JC (2000) The Influence of La Nina on African Rainfall. Int J Climatol 20:1761–1776
Nicholson SE (2009) A revised picture of the structure of the "monsoon" and land ITCZ over West Africa. Clim Dyn 32(7–8):1155–1171
Obasi GOP (2005) The impacts of ENSO in Africa. In: Sum Low P (Hrsg) Climate change and Africa. Cambridge University Press, Cambridge, 218–230
Paeth H (2010) Der Klimawandel in Afrika: Physisch-geographische Befunde und Klimamodellsimulationen. In: Glaser R, Kremb K, Drescher A (eds) Afrika. WBG, Darmstadt, pp 22–31
Rathmann J (2009) Klima- und Zirkulationsvariabilität im südhemisphärischen Afrika seit Beginn des 20. Jahrhunderts. Dissertation, Universität Augsburg

UNCCD (1994) United Nations Convention to Combat Desertification in Countries Experiencing Serious

UNCCD (2009) "Land Day" on June 6, 2009, a side event of the UNFCCC's "Bonn Climate Change Talks" (June 01–12, 2009). http://www.unccd.int/publicinfo/landday/docs/1TengbergUNDP.pdf

UNCCD "COP (11)/5" on September 27, 2013, Consideration of the follow-up to the outcomes of the United Nations Conference on Sustainable Development relevant to the United Nations Convention to Combat Desertification (September 16–27, 2013). http://www2.unccd.int/official-documents/cop-11-windhoek-2013/iccdcop115.pdf

UNCOD (1977a) Its causes and consequences. Compiled and edited by the Secretariat of the United Nations Conference on Desertification, Nairobi, Kenya, 29 August to 9 September 1977, Oxford, p 448

# 6 The Water Resources

From a hydrological perspective, Africa's most distinctive feature is its remarkable variability. Spatially, the African continent arguably embraces greater contrasts than any other continent, showcasing an unparalleled range of hydrological conditions. The seasonality and spatiotemporal variability of precipitation, as well as the first-order morphology of the continent with its ubiquitous, expansive plateau landscapes and trench systems, are responsible for Africa's distinct hydrology. Mean annual precipitation ranges from near zero over large areas of the Sahara to nearly 10,000 mm in the foothills of Mount Cameroun on the coast of the Gulf of Guinea. Likewise, mean annual evapotranspiration varies from a few millimeters in the Sahara and the Namib to more than 1,300 mm in the central areas of the Congo Basin. From a temporal perspective, variability must also be seen as a key feature of the continent's hydrology. Such variability embraces both short-term year-to-year fluctuations in annual precipitation and runoff, particularly in the drier areas where the coefficient of annual runoff is large, and longer-term cycles of dry-and-wet periods, including the Sahelian droughts.

Although some of the world's largest lakes and reservoirs are in Africa, the availability of water per capita is 2.4 times lower than the global average. Therefore, providing adequate water supply for the entire continent is already one of the central international challenges today and will likely be vital for the development of large parts of the continent in the future. Only in recent decades has the focus in Africa increasingly turned to environmental dynamics, and in the meantime, there are numerous studies and investigations on the hydrological and hydrogeological potential and its changes in space and time.

From a hydrogeographic perspective, three main regions can be distinguished in Africa. The **humid and rain forests** of the wet equatorial Central and West Africa have a sufficient year-round water supply. The **savannas and dry forests** of the wet-dry tropics and the dry tropics are characterized by a rainy season and the associated periodic abundance, respectively, seasonal lack of available water resources. The annual course of the still considerable runoff quantities shows strong variabilities with a late summer maximum. The **semi- and full deserts and thorn bush savannas** in dry subtropical Africa with less than two humid months, on the other hand, show a significant water shortage all year round.



R. Baumhauer, *Physical Geography of Africa*, Geography of the Physical Environment,
https://doi.org/10.1007/978-3-662-71406-5_6

## 6.1 Rivers and River Regimes

Both the runoff behavior and the amount of runoff from Africa's rivers are closely linked to the climatic and the first-order morphology of the continent. Thus, in the wet equatorial Central and West Africa, there are **perennial rivers** with high runoff everywhere. Rivers that flow continuously throughout the year are referred to as perennial rivers. In contrast, rivers that flow only at certain times of the year are referred to as **periodic or seasonal rivers**. Their catchment areas often extend into the wet-dry tropics and the dry tropics. The runoff is, therefore, co-determined by the annual precipitation season. Ubangi in the Congo Basin, Logone and the Chari in Cameroon and Chad, the Sanaga in Cameroon, or the lower Niger in Nigeria are typical effluents for these regions. In the rivers of the wet-dry tropical climate such as Senegal, Niger, or Zambezi in southeastern Africa, as well as the Okavango and Kunene in southwestern Africa, there are high discharge volumes with significant variabilities throughout the year and a pronounced maximum in late summer. In the dry subtropics, there are high fluctuations throughout the year and over several years. They flow partly as **exotic rivers (or diarrheic rivers)** through arid areas or large swamp areas. In this way, significant parts of the discharge are lost through evaporation or seepage, and the discharge volume of the rivers decreases significantly. In some cases, they even dry up completely or in sections, like the Limpopo at the Botswana to northern South Africa border. These arid regions are the transition areas to Wadis (or Ennedis) in the Central Sahara, Riviere, Iishana (Sing. Oshana) or Omiramba in Namibia with **episodic (or ephemeral) runoff**. In extreme cases, the flow pattern is dominated by episodic runoff that occurs only every few years or even decades. The dry subtropical climate with precipitation in winter shows temporarily high discharge volumes in winter and spring, while the riverbeds are predominantly dry in the summer months (see Fig. 6.1). An exception is the Vaal-Orange system, which receives rainfall throughout the year in its large catchment area in the highlands of South Africa.

The variability and amount of surface runoff cannot be attributed solely to differences in precipitation patterns. Other factors, such as soil crusts, including the widespread laterite crusts (Petroplinthite-) found in the wet-dry tropical regions, also play a significant role in influencing the runoff. In addition, the intensity of erosive activity also depends on whether the ground surface is covered with vegetation all year round. Because the siltation/aggradation of the exposed soil can also lead to increased surface runoff and surface flushing. In contrast to the other continents, the rivers of Africa carry only a very small sediment load. This is a result of the lack of surface runoff in the desert regions, the tectonic stability of the continent, which prevents the erosive activity of the rivers due to retrogressive erosion, as well as the dominant, erosion-resistant Precambrian basement over large areas of the continent, at least where the latter is not deeply chemically weathered. Rapids, steep lower courses of rivers or waterfalls like the **Victoria Falls** located on the Zambezi River, which forms the border between Zimbabwe and Zambia, confirm the resistant basement and restrict the navigability of the large rivers. The Congo River, with some tributaries like the Ubangi, is only well navigable in the hinterland. At the lower course, it breaks through the Lower Guinea Ridge with several rapids and drops about 300 m in the last 350 km of its flow. Even at the mouth, it is over 200 m deep in the 150 km wide shelf. The estimated mean fluvial sediment output of Africa, according to Milliman and Meade (1983) and Meybeck (1979), is 35 t per square kilometer per year (t/km$^2$/year), in Asia it is 229 t/km$^2$/year, in Europe 50 t/km$^2$/year, in North and Central America 84 t/km$^2$/year, and in South America even 589 t/km$^2$/year.

**Fig. 6.1** The flow regimes of African rivers as characterized by Walling (1996)

The large drainage systems of Africa are irregularly distributed across the continent, with the highest river density always found where the greatest water availability due to rainfall is present. This is particularly true for the Congo River and its source and tributaries. **Endorheic river systems** are rather characteristic of the African continent. They drain into inland lakes or so-called terminal lakes (for example, Lake Chad, Lake Turkana, Lake Nakuru, Lake Naivasha and Lake Manyara in East Africa, Okavango Delta with Makgadikgadi pans, and Lake Ngami) and do not form a contiguous drainage system. Occasionally, watersheds are in palaeovalleys, such as the Molopo in the South Kalahari, whose valley runs towards the Orange River but no longer flows above ground today. About a third of Africa's land area is covered by endorheic basins, including the Sahara, the Namib, the Kalahari, as well as the Chad and Turkana Lake Basin. The annual discharge of the endorheic rivers amounts to 158 cubic kilometers or about four percent of the **exorheic discharge**, i.e., the drainage into the oceans. This comprises around 3,720 km$^3$, of which 90 per cent drains into the Atlantic and only ten percent into the Indian Ocean. Except for the northward flowing Nile, most of Africa's largest rivers run latitude parallel. While the Nile flows into the Mediterranean, the Congo River and Orange-Vaal River, with their source areas in the highlands of Lesotho and Eswatini, drain over a length of 2,100 km into the Atlantic. The Niger River, with a catchment area that also includes the plateaus of the Upper Guinea ridge in Guinea, drains after 4,160 km into the Atlantic. The Zambezi River, with a catchment area on the southern slope of the Lunda Ridge, drains 2,660 km into the Indian Ocean. Large seasonal differences in discharge behavior are typical for the African rivers, especially for those with catchment areas in the savannas like Zambezi, Orange, and Niger Rivers. Only the Congo River flows predominantly through regions with year-round rainfall. It accounts for 38 per cent of the total African water volume and has, after the Amazon, the highest discharge rate worldwide (Figs. 6.1 and 6.2).

The **Nile** (6,671 km) is the longest river in the world. It flows as an allogenous (exotic) river almost exclusively through arid areas, thus showing high evaporation rates with a relatively low discharge volume (see Table 6.1). The **White Nile**, with its source arms in the East African countries of Rwanda and Burundi, is fed from the equatorial rainfall areas of Central Africa. The **Blue Nile**, from the moist Ethiopian Highlands, allows the Nile to continue north of Khartoum as an alien river with its river oasis. The approximately 22, 000 km$^2$ large delta of the Nile represents the most important agricultural economic area of Egypt. Already in 1899–1902, a large dam was built at the first cataract, and from 1960 to 1971, the **Aswan High Dam** (Sadd

**Fig. 6.2** Container ships on the Congo River at Kinshasa, DR Congo, and Brazzaville, Rep. Congo. *Photo* F. Plock, Mauritius

**Table 6.1** The flow regimes of African rivers as characterized by Walling (1996)

| River | Length of river [km] | Basin area [Mio. $km^2$] | Mean annual runoff at mouth [$m^3/s$] | Utilization (in addition to fishing and regional transport) |
|---|---|---|---|---|
| **Nile** | 6,650 | 3 25 | 140 | Irrigation, production of electrical energy from hydropower |
| **Niger** | 4,185 | 2.26 | 6.000 | Post-rainy season cultivation, seasonal pasturing, irrigation, production of electrical energy from hydropower (Kainji Dam in Nigeria) |
| **Senegal** | 1,086 | 0.37 | 647 (water level gauge Dagana) | See Niger |
| **Congo** | 4,374 | 3.78 | 40,250 (nearby Kinshasa) | Production of electrical energy from hydropower (Inga Dam) |
| **Zambezi** | 2,574 | 1.33 | 4,200 | Post-rainy season cultivation, irrigation, production of electrical energy from hydropower (Kariba Dam and Cahora Bassa Dam) |
| **Oranje** | 2,160 | 1.0 | 158 (water level gauge Irene) | See Zambezi, add. water transfer |
| **Rheine** | 1,233 | 0.22 | 2,900 (river mouth) | |
| **Mississippi** | 3,778 | 2.98 | 18,400 | |
| **Amazone** | 6,400 | 5.96 | 206,000 | |

el Ali) was built just seven kilometers upstream, damming the Nile into the **Nasser Reservoir**, the third largest in Africa. This reservoir was intended to increase the irrigation areas and secure the general water supply. The dam was also supposed to regulate the annual floods and secure Egypt's power supply with an energy output of around 10 billion kWh per year. On the other hand, due to the lack of sediment input, coastal erosion is increasing in the Nile Delta. In addition, artificial fertilizer must replace the absence of the annual Nile flood. Excessive irrigation, inadequate drainage of the saline Nile water, and rising groundwater levels also lead to increasing salinization of agricultural land. A treaty on the division of Nile water between the countries of Ethiopia (Blue Nile, Atbara), Uganda (Victoria Nile), and the Democratic Republic of Congo (Albert Nile), as well as Sudan and Egypt as the main users, is still pending. The construction of the Grand Ethiopian Renaissance Dam, which dams the Blue Nile in Ethiopia just 10 km from the Sudanese border and with 74 billion cubic meters of storage capacity is also one of the largest on the continent, has already led to political tensions between Ethiopia on the one hand and Sudan and especially Egypt on the other (see Excursus: Grand Ethiopian Renaissance Dam).

The **Congo River** is, as mentioned, the most abundant water river in Africa. The Lualaba-Congo River catchment area extends over the entire Congo Basin and is, next to the Amazon, the largest equatorial river system on earth (see Table 6.1). With a length of 4,374 km, the Congo is the second-longest river in Africa after the Nile. Measured by its mean runoff of 41,800 $m^2$ per second in the estuary area (for comparison: the Mississippi carries 18,400 $m^3$ per second in the estuary area, the Rhine at the German-Dutch border carries 2,220 cubic meters per second), it is the most water-rich river of the continent and with 220 m depth in its estuary area the deepest river on earth. Its erosion base is thus today below sea level, which suggests that the erosion was set to a (glacial) lower sea level. The catchment area of the **Lualaba-Congo** is around 3,789,000 $km^2$ after that of the Amazon, the second largest on earth, and is predominantly in the Democratic Republic of Congo. Furthermore, it extends into the Central African Republic, Angola, the Republic of Congo, Tanzania, Zambia, Cameroon, and Gabon, and includes parts of Burundi and Rwanda. The Congo is a first-rate shipping route with outstanding importance for traffic within the Congo Basin, especially for the Democratic Republic of Congo, where there are still hardly any roads or railway connections for long-distance traffic. However, due to the many cataracts, only 3,000 km of the river are navigable in sections, and direct, navigable access to the Atlantic is interrupted by the Lower Guinea Ridge. The sometimes long, non-navigable sections are bypassed by the railway. Only navigable in sections are also Ubangi and Kasai,, the two major tributaries of the Congo River. Responsible for the many cascades and rapids are the fluvially very hard-to-erode rocks of the Precambrian basement ridges and the dominant chemical weathering in the tropics. The combination of these two factors results in a significantly lower downward erosion of the Congo River and its tributaries compared to other climate zones and/or with different underlying geology. In the savanna areas of the ridges, the water level variability of the rivers between rainy and dry seasons additionally limits the use.

The **Niger River** originates on the plateaus of the Upper Guinea Ridge in Guinea, crosses in a large, northeastward arc the northern Sudan and southern Sahel zone, and flows with a delta in the lowlands of Nigeria (Biafra) into the Gulf of Guinea. Of particular interest is the **Niger Inland Delta** in Mali. It is referred to as Massina and comprises a large, seasonally flooded area of the size of Denmark between the Malian cities of Mopti and Timbuktu. Here, as in the entire river course, the river breathing, which can amount to up to ten meters in the river course of the year, has a detrimental effect on shipping, while it creates favorable conditions for post-rainy season cultivation of millet, rice, and vegetables, but also for seasonal grazing. The Niger River, with its year-round water flow, forms an important development line in semi-arid West Africa (Fig. 6.3). The construction of the Kainji Dam in western Nigeria and the oil and gas extraction in the Niger River delta and on the shelf have led and continue to lead to strong interferences in the river and coastal ecosystems. They are now largely irreversibly destroyed.

The **Zambezi River** breaks through, as already mentioned, the South African Great Escarpment and flows into the Indian Ocean. Numerous water gaps and waterfall or cataract sections, such as the 110 m deep Victoria Falls, hinder shipping, which is only of importance in sections. The Zambezi is important for energy production through two of the largest reservoirs and power plant facilities in Africa, the **Kariba Dam** on the border between Zimbabwe and Zambia, and the **Cahora Bassa Reservoir** in Mozambique (Fig. 6.4).

As an alien river whose headwater region is located at over 2,000 m altitude in the rain-rich highlands of Lesotho, the **Orange River** crosses the semi-arid and arid subtropical South Africa. Especially its tributary Vaal River, whose field of source is close to the border with Eswatini, is of great importance for the drinking

**Fig. 6.3** Niger River end of February 2006 at Ségou in Mali. *Photo* R. Baumhauer

and utility water supply of the highly developed industrial region around Johannesburg and the intensively agriculturally used irrigation areas in the provinces of Free State and Mpumalanga. Due to the large water demand of the industrial-urban agglomerations and irrigation agriculture, a "water transfer" from the humid eastern side of the South African Great Escarpment in KwaZulu-Natal to the Vaal River (the so-called Tugela-Vaal-Scheme) and since the end of the 90s of the twentieth century from the highlands of Lesotho also to the Vaal River (see also Fig. 6.5).

## 6.2 Lakes and Wetlands

Compared to other continents, Africa has only a few natural lakes. They cover a total area of 200,000 km$^2$ and, like the rivers of Africa, are characterized by a variety of lake types. In addition to the deep tectonic lakes of the East African Rift (for example, Lake Malawi or Lake Tanganyika), the lakes formed by volcanic activity (for example, Lake Kivu or Lake Chala), the shallow floodplains (for example, Okavango) and the basins (for example, Lake Chad, Kalahari), there are also some glacial lakes at high altitudes. In terms of area, **Lake Victoria**, with 68,800 km$^2$, is the third largest lake worldwide. However, when considering the water volume, **Lake Tanganyika** is larger. Due to its location in the Tanganyika Lake Rift (Rift zone), it has an enormous water depth of 1,470 m, reaching almost 700 m below sea level, and contains 17 per cent of the Earth's surface drinking water. In addition to these freshwater lakes, there are several salty and brackish waters, which are only moderately usable. Partially, they dry up completely during the dry seasons. A special form in some regions of North Africa is the so-called **Chotts**, salt lakes that dry up periodically (Table 6.2 and Fig. 6.6).

**Fig. 6.4** Cahora Bassa Dam on the Zambezi River in Mozambique. *Photo* A. Martens, Mauritius

The **Lake Chad** and its basin hold a special position: With an area of 2.5 million square kilometers, it is the largest endorheic basin on Earth. Most of the precipitation falls between June and August in the southern third of the basin. In the remaining part, arid climate conditions dominate. Following this pronounced precipitation gradient from the southern to the northern part, about 90 per cent of the runoff from the catchment area is drained into Lake Chad via the two southern tributaries **Chari River** and **Logone River**. The remaining ten percent comes from the Komadugu Yobe, which flows into Lake Chad from the west. Due to a natural geological barrier on the lakebed, the lake is divided into two basins, which form a contiguous lake surface under favorable climatic conditions. Increasing water extraction for irrigation and decreasing rainfall in the catchment area have led to a decrease in the very shallow lake with average water depths of two meters by 90 per cent since about 1970. Longer periods of drought in the 1970s, 1980s, and since the turn of the millennium, together with high evaporation rates, have promoted drying up. Currently, only a minimal area of the original lake in its southernmost part is covered by water all year round.

Although they cover only about one percent of Africa's land area, spanning approximately 340,000 km$^2$, **wetlands** hold significant hydrological, ecological, and economic importance, providing a striking contrast to the extensive arid and semi-arid regions of the continent. Africa's wetlands are not tied to any specific climate zone. They can be found everywhere in the tropics and are predominantly concentrated near the equator due to the existing moisture. Typical are perennially or periodically high groundwater levels. Often, the flooding of the wetlands only occurs significantly after the local rainy season, as in the **Okavango Delta**, which shows enormous seasonal variabilities. Perennial wetlands show high moisture losses through evaporation, which can be higher than the actual surface runoff. The wetlands can be divided into

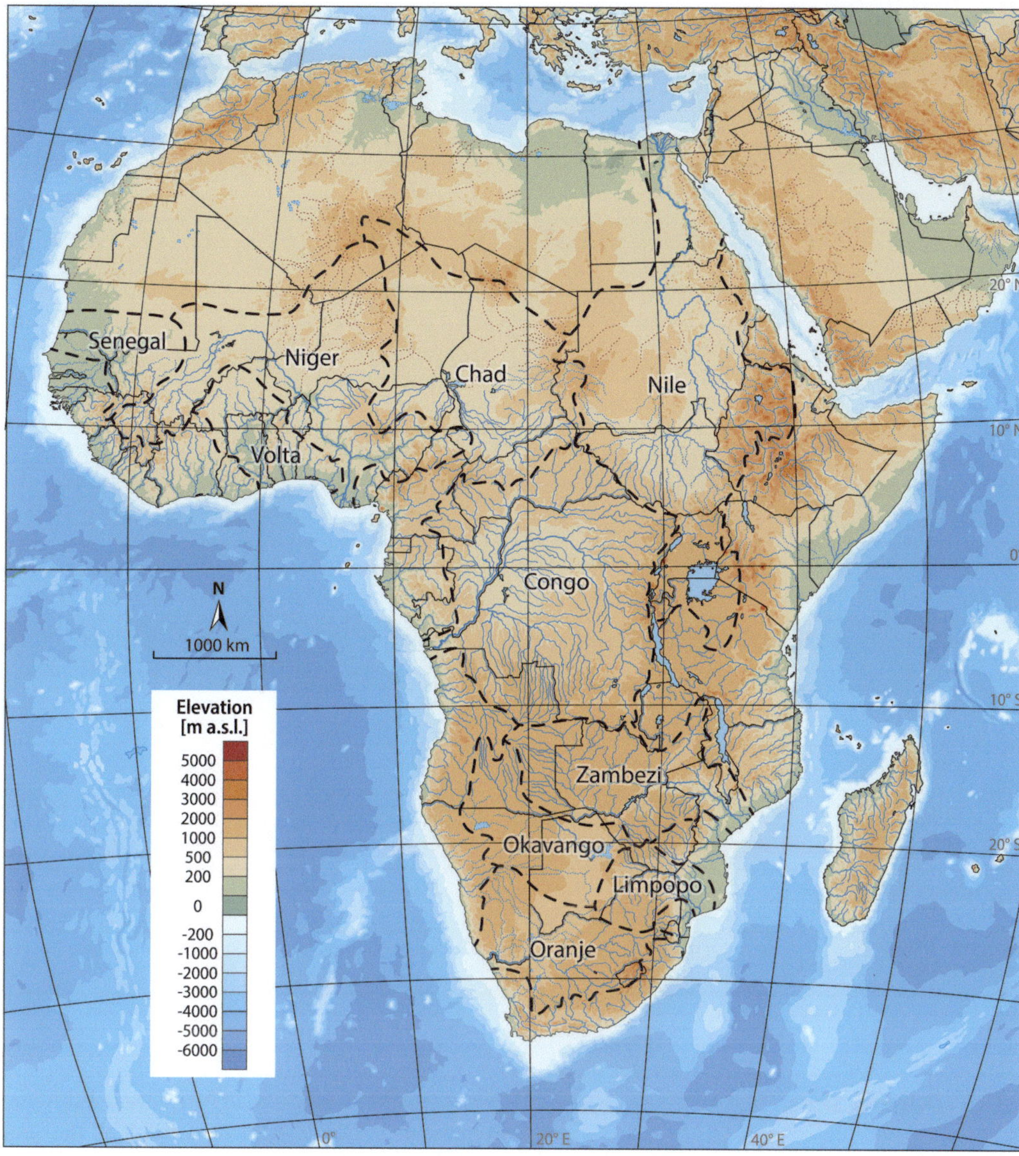


**Fig. 6.5** The drainage basins of the largest rivers in Africa and the endorheic basins of Lake Chad, and Okavango

five subtypes: floodplains, riparian forests, marsh landscapes, ponds, and bog forests. Wetlands in floodplain areas are located particularly between the riverbed and the valley flanks or river terraces in those areas that are periodically flooded by water and can cover a wide range of sizes and extents: narrow strips like on the Kenyan river Tana or wide plains in the lower Congo Basin. Larger floodplain landscapes also include marsh landscapes, which are widely found in the peripheral areas of flat lakes. They are widespread in Africa and are often referred to as swamps. One of the largest areas of this type worldwide is the Sudd in Sudan. Other large swamp areas are located at the Okavango River, at Lake Kyoga in Uganda, and in the Bangweulu Wetlands (-Swamps), but also the Luena Flats on the upper Zambezi River or the Kafue Flats, both in Zambia. The largest peatland on earth was discovered in the central Congo Basin in 2017, the Cuvette Centrale.

**Table 6.2** The dimensions of large natural and anthropogenic lakes in Africa

| Lake | Bordering states | Area [km²] | Max. depth [m] |
|---|---|---|---|
| Lake Victoria | Tanzania/Uganda/Kenya | 68,800 | 85 |
| Lake Tanganyika | Tanzania/D.R.Kongo/Zambia | 32,893 | 1,470 |
| Lake Malawi | Malawi/Mozambique | 29,600 | 704 |
| Lake Chad | Chad/Nigeria/Niger | 10/25,000 (min/max) | 4-11 |
| Lake Bangweulu | Zambia | 5,500/10,000 (min/max) | 4 |
| Lake Albert | Uganda/D.R.Kongo | 5,347 | 48 |
| Leke Mweru | Zambia/D.R.Kongo | 5,120 | 27 |
| Lake Tana | Ethiopia | 3,000 | 14 |
| *Lake Constance* | *Germany/Austria/Switzerland* | *536* | *251* |

| Anthropogenic reservoirs (area [km²]) | Bordering states | Volume [km³] | Installed hydro-electric power [MW] |
|---|---|---|---|
| Volta (8.502) | Ghana | 153 | 912 |
| Kariba (5,580) | Zambia/Zimbabwe | 180.6 | 1,320 |
| Lake Nasser/Assuan (5,248) | Ägypten | 165 | 2,100 |
| Cahora Bassa (2,800) | Mozambique | 65 | 2,075 |
| Grand-Ethiopian Renaissance Dam (1,874) | Ethiopia | 74 | 6,000 |

## 6.3 Dams/Lakes

Until the mid-twentieth century, there were no reservoirs in Africa. Between 1954 and 1977, four of the five largest reservoirs in the world were built in Africa: The **Kariba Dam** on the Zambezi, the **Volta Reservoir** on the Volta River, the **Kainji Reservoir** on the Niger, and the **Lake Nasser** on the Nile. In 2022, the **Grand Ethiopian Renaissance Dam** in Ethiopia, one of the largest reservoirs in Africa, was completed (see Excursus: Grand Ethiopian Renaissance Dam). Almost a third of the total African runoff is stored in the reservoirs of the continent, thus the largest proportion of runoff compared to other continents. Currently, more than 100 larger reservoirs, with a total storage capacity of over 1,000 million cubic meters volume, alter the runoff patterns of the dammed rivers due to water use and regulation. As a result of the larger water surface area, there is increased evaporation.

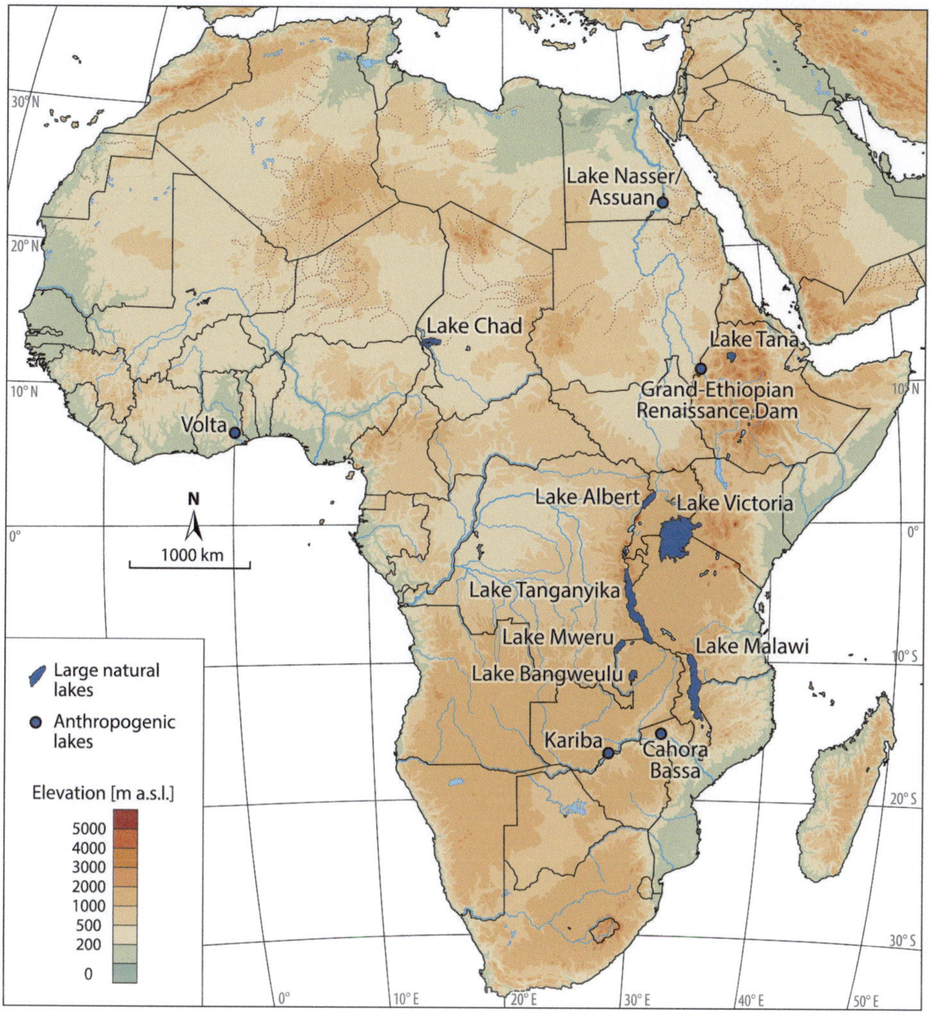


**Fig. 6.6** Major natural lakes and anthropogenic reservoirs in Africa

### Excursus: Grand Ethiopian Renaissance Dam—The Struggle for the Water of the Nile

The Grand Ethiopian Renaissance Dam (GERD) has been under construction since 2011 and was completed in 2022. The dam project, with a nearly two-kilometer-long and 145 m high main gravity dam, dams the Blue Nile in the remote western Ethiopian region of Benishangul-Gumuz and is only 10 km east of the Sudanese border. The storage capacity is 74 billion cubic meters, and the attached hydroelectric power plant will be the largest in Africa with 6,000 MW. However, the reservoir also represents a high-risk bet on more abundant annual rainfall than is regularly measured in the catchment area of the Blue Nile due to considerable rainfall variability. A large dam can compensate for this through multi-year storage, but it is expected that instead of the planned 6,000 MW, only about 3,000 MW will be usable on average throughout the year at constant output. The main benefit for Ethiopia will be the production of electrical energy from hydropower and the stabilization of the availability of electricity in the power grid.

The political dimension of the project is more problematic. The Nile is the lifeline of Ethiopia, the Republic of Sudan, and especially Egypt. Egypt, which has been against the construction of the dam from the beginning, sees its existence threatened. Although the Ethiopian government has always argued that the dam would not reduce the availability of water downstream but only regulate it, thereby making more land available for irrigation, Egypt's political resistance continues, fearing for its water supply, especially during the multi-year flooding of the lake. During the filling phase of the reservoir, Ethiopia holds back a lot of water. Despite Egypt's Aswan Dam, hitherto Africa's largest hydroelectric power plant, will lose about 12 per cent of its power during the years of filling due to inflows. As early as 2016, two French engineering offices were commissioned by the three affected countries to develop a proposal on how the lake could be filled in a tolerable manner for the neighboring states. An agreement is still pending to this day. In contrast to Egypt, the NGO "International Crisis Group," among others, does not see the water supply of Egypt's 100 million inhabitants threatened by mean annual rainfall due to the regulating effect of the Aswan reservoir. However, the government in Cairo refers to a treaty from 1929. In this treaty, Egypt was promised a significant portion of the Nile water and a veto right for upstream constructions. In a second agreement from 1959, Egypt was allocated about 66 per cent of the river water, and the Republic of Sudan, founded in 1956, about 22 per cent. Ethiopia was not part of these agreements. In 2010, the bordering countries upstream of the Nile concluded their treaty without including Egypt and the Republic of Sudan, the so-called Framework Agreement for Cooperation. According to this, construction projects are allowed even without approval from Cairo.

In March 2015, the heads of state of Ethiopia, the Republic of Sudan, and Egypt signed a declaration of principles in Khartoum, which included studies on the after-effects of the dam, rules for filling the reservoir, an agreement on annual discharge quantities, and procedures for resolving disputes. Ethiopia wants to base this on mean rainfall; Egypt and Sudan, on the other hand, are pushing for provisions for dry years. Further contract negotiations between the three states in June 2020 failed. The Republic of Sudan, which fears effects on its downstream Roseires Dam and sees the fate of millions of people at risk, then called on the UN Security Council and under the leadership of the African Union, direct talks between the conflicting parties were resumed at the end of June 2020. However, after Ethiopia began damming the Blue Nile in August 2020, Egypt, which draws 90 per cent of its usable water from the Nile, left the negotiations. In February 2021, the Republic of Sudan again called for international mediation. In response, the foreign ministers of Egypt, Ethiopia, and Sudan met in Kinshasa in April 2021 for "final" talks, which also ended without result.

## 6.4 Groundwater

Fifty to seventy-five percent of the African population is directly dependent on groundwater resources. Groundwater is essential for drinking water supply and is also increasingly used for irrigation of agricultural areas. In general, groundwater occurrences are strongly dependent on the infiltration of precipitation into the soil and the infiltration from the bank area of surface waters into the subsoil, as well as the geologic setup and storage capacity. Accordingly, the groundwater resources on the African continent are unevenly distributed (Fig. 6.7).

Volcanic rocks and the rocks of the Precambrian basement have only a low

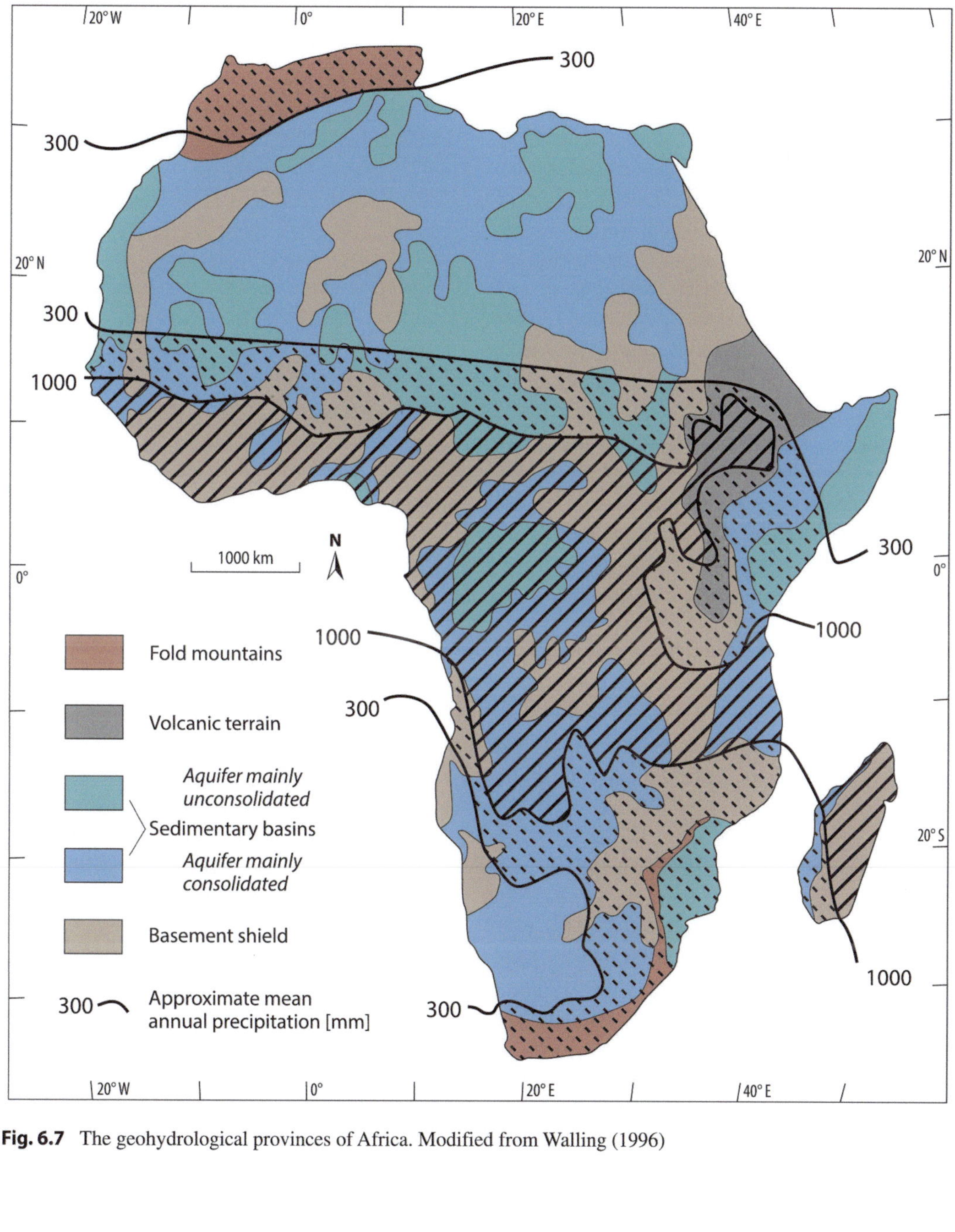


**Fig. 6.7** The geohydrological provinces of Africa. Modified from Walling (1996)

water-holding capacity. In contrast, the storage capacity of the sedimentary rocks of the large basins is very high. They form the most important groundwater reservoirs of the continent. The total groundwater reserves in Africa are estimated to be 0.66 million cubic kilometers. Thus, they are more than 100 times the annually renewable freshwater resources, more than 20 times the freshwater stored in Africa's lakes, and about 30 times as much water as falls as precipitation across the entire continent each year. The larger groundwater reservoirs are in the sedimentary Saharan aquifers of Libya, Algeria, Egypt, Sudan, and Chad. Despite the lack of surface water, groundwater deposits are present at varying depths in almost all natural desert areas of Africa, sometimes on several levels. However, most of this groundwater is a fossil and will no longer be recharged under current climate conditions. In North Africa, the Nubian Sandstone is one of the most important aquifers. A strategy paper by a US think tank on food security in Africa noticed: "The Eastern Erg in Algeria and Tunisia covers an area of almost 400,000 km$^2$ and stores an amount of water equal to about four times the average annual renewable supply of the entire North Africa and Middle Eastern region. Only 0.04 per cent of this volume is recharged annually, so this is essentially fossil water.

The Nubian Sandstone Aquifer underlies parts of Egypt, Libya, and Sudan, extending over an area of 1.8 million square kilometers. The volume of stored water is nearly 20 times the average annual renewable supply for North Africa and the Middle East, and the aquifer has an annual recharge rate equal to about 2.5 per cent of its volume" (Rosegrant and Perez 1997, p. 38).

In the future, a reliable and increased water supply in Africa will largely depend on the availability of groundwater, which, however, is only guaranteed in the humid regions of Central, East, and West Africa. Fossil, non-rechargeable groundwater is already being used for irrigation today. It is expected that the use of groundwater for irrigation will increase significantly, not least to counter growing food insecurity. In many urban agglomerations, drinking water supply can already only be ensured through deep wells.

### Excursus: Problems of Groundwater Development and Use in Arid Areas Using the Example of the Sahara Oasis Bilma in the Republic of Niger

Artesian groundwater bodies with fossil freshwater and near-surface, highly saline deposits determine the groundwater conditions in the central Saharan oasis of Bilma. The coexistence of the two groundwater levels indicates a complicated, small-chambered hydrogeological system associated with tectonic activities. Artesian waters flow freely from numerous spring hills, some reaching up to 15 m high, providing a vital source of water for the oasis population. Non-evaporating surplus water from these natural artesian springs flows to the near-surface groundwater horizon. In addition to its location in saline loose sediments, an influx from leaky artesian levels also led to the creation of one of the largest salt pans in the sub-Saharan region.

Already in the twelfth century, the salt pans of Bilma were mentioned by the Arab geographer Muhammad al-Idrisi (around 1100–1166), and Herodotus reported about salt houses and salt production in this area of the Sahara about 1,500 years earlier. For a long time, therefore, there has been a balance between an artesian groundwater reservoir that formed in the Holocene, whose water was formed according to 14-C age dating 5,500–6,000 years ago, and a free, locally limited groundwater body with high salt contents. These water reserves served the oasis population for modest gardening, which, along with the exploitation of the salt pans, became the actual "survival factor" (economic factor) for the population of Bilma. However, the occurrence has also significantly decreased due to the increasingly arid climate conditions of the Late Holocene and consequently decreasing groundwater recharge rates since the formation period.

**Fig. 6.8** Uncontrolled outflow of fossil groundwater in the Central Saharan oasis of Bilma, Rep. of Niger, after a failed drilling (see Excursus: Problems of Groundwater Development). *Photo*: R. Baumhauer

This balance was disturbed by two deep wells drilled in 1980 and 1981. Lack of knowledge of exact (hydro) geological conditions as well as technical inadequacies led to the uncontrolled leakage of one of the wells (Fig. 6.8). As a result, since 2006, in addition to the already occurred salinization of large areas of the main oasis Bilma and the rapid loss of storage pressure of the fossil groundwaters, not only the water supply of the population from this drilling is no longer guaranteed, but also there has been an increasing impairment of salt extraction in the salt pans.

The uncontrolled leakage of high-quality fossil water creates numerous problems for the oasis Bilma. Three aspects are highlighted here.

Already occurred is the formation of spatially and temporally strongly differing open water surfaces in the deeper northern and eastern parts of the oasis. In this area, numerous gardens had to be abandoned due to increasing salinization. The constant freshwater inflow into the lowest parts of the oasis, where the salt pans are located, results in a reduction of the total content of the near-surface saline groundwater there in dissolved substances, thus, a dilution effect can be recorded. Today, the sensitive natural balance between the small amounts of incoming freshwater and the salt-rich sediments, which form the basis for optimal salt extraction, is increasingly disturbed. At least in the medium term, there is a risk that the essential economic factor of the region and, thus, a central part of the livelihoods will be destroyed. If you also include the Tuareg of the Aïr Mountains, who live from salt transport and trade, the negative effects of anthropogenic changes increase considerably. The estimated annual production of salt before 1906 was 10,000 t (Gadel 1907), 1953–1960 was still about 1,000 t annually (Faure 1963; newer numbers are not available). The salt production includes both table salt (so-called Beza salt) and livestock salt (so-called Kantou salt), which makes up 70 per cent of the production and is of great importance as a lick for livestock farming in the Sudan zone. The traditional salt trade by Tuareg salt caravans (Azelei) still works today in the form of a triangular trade:

1. Agadez/Aïr Mountains—Bilma: Dates, millet, and consumer goods
2. Bilma—Nigeria: Salt
3. Nigeria—Agadez/Aïr Mountains: Millet and consumer goods.

Attempts in the 1980s by, among others, the GTZ (the predecessor of GIZ, the German Society for International Cooperation is an organization for development cooperation, which operates internationally on behalf of the Federal Republic of Germany) to shift the salt trade to trucks, failed for both political and economic reasons.

## 6.5 Challenges

Africa's water resources are, in the future and already in large parts of the continent today, characterized by scarcity and poor quality, especially due to pollution. Due to the seasonality and spatiotemporal variability of precipitation and, more fundamentally, the increasing risk of drought due to climate change, water scarcity is particularly evident in the arid and semi-arid regions of the continent as a major daily nutrition and development problem. Growing population numbers enhance this. Over 40 per cent of the population south of the Sahara is not adequately supplied with clean water. In addition, the disparities between rural and urban areas, both in terms of necessary water transfer and water distribution into the respective regional and local sub-areas, as well as between rich and poor, are considerable. Population growth and increased urbanization also result in increased water pollution, promoted by economic developments without sustainable ecological harmony, such as the increased use of fertilizers and other harmful chemicals. Pollutions of this kind, as well as land use changes, the introduction of foreign, invasive animal and plant species or other anthropogenic influences, such as soil degradation, lead to far-reaching impacts on the natural environment and ultimately to its destruction. While Africa's water demand will continue to rise in the future, global climate change is expected to exacerbate this situation. Against the background of a predicted 1.5 times stronger warming of Africa compared to the global expected, the resulting impacts on the availability of drinking water are still completely unknown and difficult to predict. However, it should also be noted that water availability is not least dependent on political and institutional problems. In addition to political will, there is often also a lack of capital to carry out investments in infrastructure and organization. Therefore, a comprehensive reform of water policy involving all social and economic actors as well as the institutional and legal environment is seen as essential. Water saving and associated productivity increase, as well as environmentally friendly and, above all, sustainable water use, are of crucial importance. Accordingly, the challenges resulting from increased water scarcity can be addressed by three sustainability strategies:

- Supply management in the sense of localization, development, and use of new water sources (consistency).

- Demand management with the promotion of water protection and thrift (sufficiency).
- Efficient water use (efficiency).

Required for such challenges is, on the one hand, a comprehensive understanding of Africa's climate variability to estimate the effects of future climate and thus environmental changes, and on the other hand, a comprehensive investigation of the groundwater storage to determine their properties, recharge, and sustainable extraction rates. Of particular interest are those aquifers that have a sufficiently high recharge rate. Strategies to increase water supply or irrigation with the aim of high yields, which are predominantly pursued by rapidly growing cities, are very likely to fail with limited resources. Rather, the focus should be on water use efficiency, especially on avoiding unnecessary water losses. Efficiency in sub-Saharan Africa is quite low at 20–30 per cent. Only in North Africa, for example, in Morocco, it is slightly higher at 40–45 per cent and could be significantly increased by the renewal of infrastructure. Future improvements also depend on national political and economic conditions.

## References


Faure H (1963) Inventaires des évaporites du Niger (Mission 1963). Min. Trav. Publ., des Mines et de l'Hydraulique, Niamey, Niger, p 162

Gadel M (1907) Notes sur Bilma et les oasis environnantes. Rév Col 57:361–386

Meybeck M (1979) Concentrations des eaux fluviales en éléments majeurs et apports en solution aux oceans. Rev de Géologie Dyn et de Géogr Physique 21:215–246

Milliman JD, Meade RH (1983) World-wide delivery of river sediment to the Oceans. J Geol 91:1–21

Rosegrant MW, Perez ND (1997) Water resources development in Africa: a review and synthesis of issues, potentials, and strategies for the future. EPTD discussion papers 28. International Food Policy Research Institute (IFPRI), p 38

Walling DE (1996) Hydrology and rivers. In: Adams WM, Goudie AS, Orme AR (eds) The physical geography of Africa. Oxford University Press, New York, pp 103–121


## Further Readings


Adams WM (1996) Lakes. In: Adams WM, Goudie AS, Orme AR (eds) The physical geography of Africa. Oxford University Press, New York, pp 122–133

Allan JA (ed) (2013) Handbook of land and water grabs in Africa: foreign direct investment and food and water security. Routledge, London, p 488

Balek J (1977) Hydrology and water resources in tropical Africa. Elsevier, Amsterdam

Coe MT, Foley JA (2001) Human and natural impacts on the water resources of the Lake Chad Basin. J Geophys Res 106(D4):3349–3356

Conway D, Persechino A, Ardoin-Bardin S, Hamandawana H, Dieulin C, Mahé G (2009) Rainfall and water resources variability in Sub-Saharan Africa during the twentieth century. J Hydrometeorol 10(1):41–59

Faure H (1966) Reconnaissance geologique des formations sedimentaires postpaleozoiques du Niger oriental. Mém. 13. B.R.G.M. 47, p 630

Gao H, Bohn TJ, Podest E, McDonald KC, Lettenmaier DP (2011) On the causes of the shrinking of Lake Chad. Environ Res Lett 6:1–7

Hughes FMR (1996) Wetlands. In: Adams WM, Goudie AS, Orme AR (eds) The physical geography of Africa. Oxford University Press, New York, pp 267–286

MacDonald AM, Bonsor HC, Dochartaigh BÉÓ, Taylor RG (2012) Quantitative maps of groundwater resources in Africa. Environ Res Lett 7(2):1–7

Muchie M, Baskaran A (eds) (2017) Solutions to access safe and drinking water in Africa. Africa World Press, p 240

Scheren P, Diop S, Niang A (eds) (2021) Climate change and water resources in Africa: perspectives and solutions towards an imminent water crisis. Springer, Heidelberg, Berlin, p 521

Shahin M (2002) Hydrology and water resources of Africa. Kluwer Academic Publishers, Dordrecht, p 659

Shiklomanov IA (2009) Regional distribution of rivers and streams in Africa. In: Dooge JCI (ed) Fresh surface water. EOLSS Publishers, Oxford, pp 230–260

Taylor RG, Koussis AD, Tindimugaya C (2009) Groundwater and climate in Africa—a review. Hydrol Sci J 54(4):655–664

UNEP (2006) Africa's lakes: atlas of our changing environment. UNEP, United Kingdom, p 90

Wright EP (1992) The hydrogeology of crystalline basement aquifers in Africa. Geol Soc Spec Pub 66:1–27

# 7 Soils, Soil Erosion, and Desertification

The soil is arguably Africa's most important resource, serving as the foundation of the predominantly agrarian economies that characterize most countries on the continent. Starting from the most important soil-forming factor, the climate, the spatial distribution of the different **(zonal) soil types** in Africa is oriented along the latitudinal climatic zones. Only in East Africa do the smaller-scale mosaics of relief, climate, and parent material (rock) lead to regional and local differentiations, resulting in a uniquely wider distribution of **azonal soils**. Since large mountain ranges are missing and altitudes above the vegetation limit only occur very punctually, the climate-soil relationships in Africa are generally not under the influence of altitude-soil relationships. Developed soils cover around 30 million square kilometers and thus approximately 71 per cent of the African land surface. 20 per cent of the continent's area comprises deserts with loose substrates and, at most, paleosols. Nine percent are occupied by rocks directly exposed at the surface, water or snow surfaces.

Examining the zonal soils of Africa, it becomes clear that soil formation in the wet equatorial tropics, where annual precipitation exceeds 1,500 mm, is dominated by chemical weathering processes. These soils are characterized by deep-reaching soil formation, often extending several tens of meters downward, and are dominated by two-layer clay minerals (predominantly kaolinite). Additionally, the formation of iron and aluminum oxides is a notable feature, with proportions increasing in response to more intense and prolonged hot-humid conditions. Their presence is therefore considered an important environmental indicator. Other important soil-forming processes are those that occur through oxidation of iron, as well as **rubefaction**, the typical soil-forming process in wet-dry tropical climates with decalcification and hematite formation. Often associated with this is **desilication** in an at least seasonally very humid, mostly acidic soil environment, as well as bioturbation, primarily by insects and burrowers. Resulting from the high amounts of water falling and the associated strong solution, lixiviate, and relocation processes, the soils are nutrient-poor. Exceptions are alluvial soils and soils on volcanic rocks, but also those that are ameliorated by seasonal nutrient input from the subtropics. Weathering and mineral formation, as well as organism activity and litter decomposition, are intense due to high temperatures and strong soil moisture. New nutrients can be supplied in undisturbed conditions through the decomposition and mineralization of a pronounced vegetation cover. As a result of interventions or anthropogenic soil use, the soils impoverish and degrade quite quickly. In the wet-dry tropics, on the other hand, the **ferralitization** with destruction of the silicates by **hydrolysis**, intensive leaching of Si, Ca, Mg, K, Na, relative enrichment of sesquioxide's and



R. Baumhauer, *Physical Geography of Africa*, Geography of the Physical Environment,
https://doi.org/10.1007/978-3-662-71406-5_7

weathering-resistant oxides such as zircon, tourmaline, rutile, as well as the new formation of two-layer clay minerals, for example, kaolinite, is the essential soil-forming process, but mainly active during the wet and transitional periods. In regions with 1,000–1,500 mm annual rainfall and less than seven humid months, the soil-forming processes of **lessivation** (relocation of clay mineral particles into deeper soil horizons) and the **browning** (oxidation of iron compounds of iron-containing minerals) predominate. The moisture change between dry and rainy seasons influences soil development. In the case of highly clayey, often calcareous sediments or basalts predominantly in valley bottoms, depressions, and at the foot of slopes, **Vertisols**, dark, deep, and clay-rich soils develop. In the regions of Africa with less than 1,000 mm rainfall and less than four humid months, the physical weathering with **carbonatization** and often also **salinization**. The low amount of precipitation and the strongly inhibited infiltration of water due to high evapotranspiration hinder soil development. Nevertheless, weathering products are formed, which in the solum (upper humus-, fine earth- and nutrient-rich soil part) are enriched in the form of easily soluble minerals (Fig. 7.1).

**Excursus: Soil-Forming Factors and Processes**

The climate is the most important soil-forming factor, especially temperature and precipitation, which influence soil formation directly or indirectly via vegetation and water balance, and control the soil-forming processes. The most important soil-forming processes, such as ferralitization, rubefaction, calcification, salinization, alkalization, and gleization, are directly dependent on the respective climatic conditions. In Africa, therefore, the soil zones largely correspond to the latitudinal climatic zones. These zonal soils differ from the azonal soils, which are less developed. Their genesis is significantly influenced by more locally acting soil-forming factors such as parent material (rock, lithology), relief, or soil moisture balance. Depending on the parent rock and its mineralogical composition, there are clear differences in soil properties and soil manifestations. The relief position (topography) determines thickness, structure, soil skeleton content, soil moisture, the presence of hardened horizons or iron concretions, as well as color, clay minerals, humus and nutrient content, and the proportion of weatherable minerals. These factors each depend on the respective relief position, that is, whether the soil profile is at the peak or ridge, upper slope, middle slope, lower slope, valley floor, etc. The apparent correlation between the distribution of soils and vegetation in Africa serves as a natural protective mechanism against erosion. Intrazonal soils deviate from the typical situation of the climatic zone due to their water or salt balance, for example, hydromorphic soils or salt and alkali soils.

## 7.1 The Major Soil Groupings of Africa

The following describes the major soil groupings of Africa, their properties and distribution, and their share of the total area of the soil-covered regions of Africa. The basis for the identification of the soil types is the climatic zones (regions) of Africa on the one hand and the reference soil groups of the **World Reference Base for Soil Resources** (WRB) of 2015 (see Excursus: Soil classification) on the other hand.

**Excursus: Soil Classifications**

Among the multitude of different soil classifications (for example, the German Soil

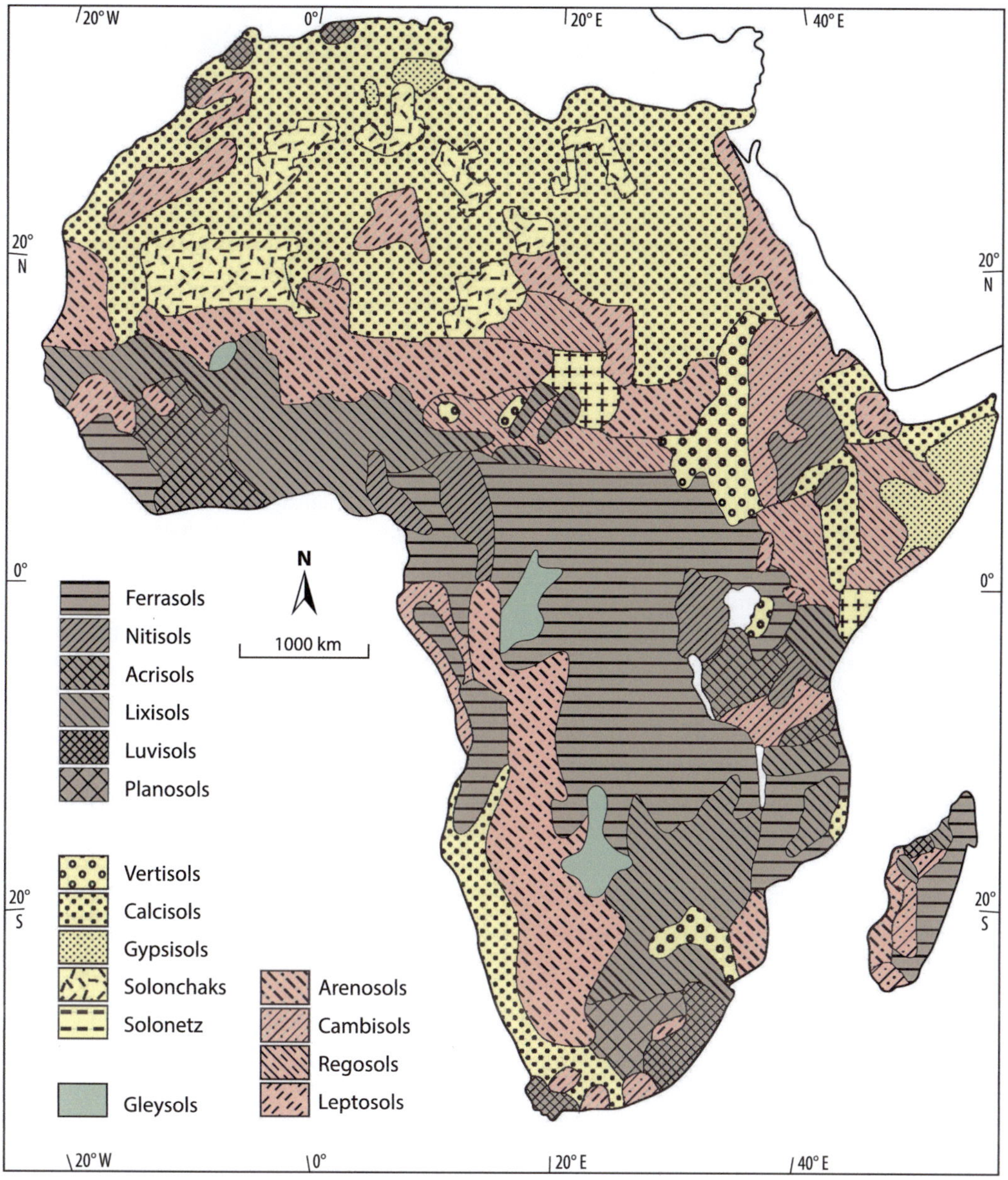


**Fig. 7.1** The major soil groupings of Africa

Science Society, DBG, or the U.S. Soil Taxonomy 1975, and many others), the World Reference Base for Soil Resources is used here. Starting in the year 1961, the FAO (Food and Agriculture Organization of the United Nations) began to develop a world soil map. The main obstacle was that the existing information could not be inserted into a generally accepted legend. In addition to the actual map creation, a legend had to be developed. The sheets at a scale of 1:5,000,000 were published in the years 1971–1981. A final revision was published in 1988. Although the legend was not an actual classification, especially not claiming to be completely consistent in considering local peculiarities, it had established itself as the "FAO classification" in the international exchange of soil science information. The highest classification criterion was the 28 "Major Soil Groups". The significantly higher number of units at the highest level of integration, compared to other classifications, is justified by its function as a map legend. Even coarser maps should still allow for meaningful differentiation (see, for example, Soil Taxonomy or even the German classification of the DBG). The nomenclature in the FAO classification tried to continue using traditional, well-established soil names (for example, Chernozem, Chestnut soil, Podzol, Solonetz, Solonchak). If such widely accepted names were not available, terms were borrowed from Soil Taxonomy. On the initiative of the FAO and UNESCO, supported among others by the International Soil Science Society, it was aimed to further develop the FAO system into an international soil classification. This classification is called "The World Reference Base for Soil Resources", or WRB for short. After several drafts and numerous international conferences, the first final version was adopted in Montpellier in 1998. This was followed by the second (2006) and third version (2015) with respective updates until 2017. With the WRB system, a world soil classification was available for the first time. The applicability is facilitated, as many elements of the FAO legend were adopted. The determination of the soil type is based on very precisely defined diagnostic elements, which are treated in the identification key separately according to horizons, properties, and parent materials. Diagnostic properties can be, for example, abrupt grain size change, permafrost or aridity. Diagnostic materials are, for example, floodplain sediments or anthropogenic materials. See World Soils Resources Report No. 84, FAO, Rome, 1998; http://www.fao.org/ag/agl/agls/wrb/Newkey.htm and https://www.fao.org/3/i3794en/I3794en.pdf.

In the wet equatorial tropics of Africa, **Ferralsols (Latosols)** are widespread, in addition Lixisols, Plinthosols and depending on the location other azonal and intrazonal soil types occur, in West Africa partly **Acrisols**, **Ferralsols** are found especially where they have been able to develop since the Tertiary under constantly humid warm conditions and protected by perennial forest areas. The warm, humid climate controls the intensity of soil formation. Deep chemical weathering by hydrolysis leads to a deep mantle of waste, which can be more than 80 m thick in some cases. The soil-forming processes of Ferralitization and Plinthisation lead to kaolinite- and sesquioxide-rich soils with a stable microstructure. Ferralsols cover about ten percent of Africa's area. These are highly weathered, leached, and acidic soils (pH values of 4.5–5.5) with low nutrient holding capacity, low base exchange capacity and base saturation, and low variance in mineral content. They are widespread in the wet tropical lowland rain forests of the central Congo Basin, on the Guinea coast, and in Madagascar. To a lesser extent, in the marginal areas of the wet equatorial tropics **Nitisols** also occur. In the floodplains and flooded areas, in addition to **Gleysols** and **Fluvisols**, **Histosols** can occur, which, together with the **Podzols**, give the rivers their color and low acidity. These characteristic soils of the tropical rainforests rapidly lose their inherent properties after mechanized clearing, a process that utilizes machinery to clear vegetation, disturbing the ground surface and uprooting woody vegetation. This process swiftly depletes the soil's limited nutrients and humus. Shifting cultivation, a traditional land-use practice, is better suited to these soils, but requires subsequent fallow periods of ten to twenty years to replenish soil fertility. More sustainable forms of agroforestry, either with simultaneous fallow with annual plants under trees and shrubs or intensive fallow with a change between arable use and fast-growing, soil-improving tree cultures that prevent erosion. However, agroforestry is hardly practiced in Africa.

**Plinthosols** are another type of deeply weathered soil, covering approximately five percent of Africa's land area. They are primarily found in transition zones between the wet equatorial and wet-dry tropics, typically in basin areas of Savannas. The essential soil-forming process is the accumulation of iron (and manganese) concretions (plinthite) in the substratum due to fluctuating groundwater levels. If they reach the earth's surface, they harden irreversibly. Over time, this creates hard, non-diggable crusts, referred to as **Petroplinthite**, which can even lead to an inversion of relief (or relief reversal). The more resistant Petroplinthite, which has formed in the substratum, is worked out by the weathering and erosion of the less resistant, overlying solum as full forms. Plinthosols are not suitable for arable farming. Only their use as road building material or bricks is possible (Fig. 7.2).

**Lixisols** cover about four percent of Africa's area. They form predominantly in the thorn tree and tall grass Savannas of the dry tropics in the Sahel, in Tanzania, Mozambique, and Madagascar. These are slightly acidic soils with a clay-enriched substratum and low nutrient holding capacity. They are dominated by kaolinite with low base exchange capacity and overall have little organic matter and a poorly developed structure. Lixisols are very susceptible to erosion by water or wind and are therefore unsuitable for the cultivation of corn, peanuts, or sweet potatoes. Lixisol sites are better used for pasture and forestry with tree cultures such as mango.

Due to relocation processes, **Alisols** (on one percent of Africa's area) have a higher clay content in the substratum (subsoil) than in the solum (topsoil). In the substratum, three-layer clay minerals (smectites) also accumulate. They have a high base exchange capacity. Alisols, which occur in the wet-dry tropics of the Sudan zone of West and East Africa, are characterized by a strongly acidic soil reaction. Mineral weathering releases aluminum, which is toxic to many organisms. Together with nutrient poverty and water stress, agricultural use on this soil (coffee, tea, sugarcane, rubber, oil palm) is strongly limited.

In the wet-dry and dry tropics, besides smaller areas with **Alisols**, mainly **Nitisols, Acrisols**, and **Cambisols** are widespread. In the summer-wet Central and East Africa with annual rainfall

**Fig. 7.2** Track in petroplinthite (formerly also referred to as laterite) in Angola. *Photo* Ibdodane, Mauritius images

of 1,000–1,500 mm, Nitisols, deep, well-developed, kaolinitic, red soils on basic and iron-rich materials, for example, basalt, diorite or gabbro are characteristic. Nitisols also have a high content of clay, iron oxides, and humus, as well as a high structural stability. They represent the best soils in terms of production ecology in the tropics and subtropics and are agriculturally used for the cultivation of coffee and cocoa. Acrisols are acidic soils that develop on acidic parent material, dominated by kaolinite and occur especially in the wetter regions of the tropics and subtropics. In the substratum, there is a clay enrichment horizon of usually two-layer clay minerals (among others kaolinite) with low base exchange capacity and a potential base saturation of less than 50 per cent. Due to their predominant nutrient poverty, larger fertilizer inputs are needed for good agricultural productivity, such as in the cultivation of coffee or oil palm.

If the annual rainfall decreases to less than seven humid months and below 1,000 mm, then on low-calcium and silicate parent rocks, **Cambisols** develop (on 11 per cent of the area of Africa). These are young soils with clearly recognizable, but not far-advanced substratum development and correspondingly slight horizon differentiation. The pH value in the solum (topsoil) is 5.0–6.5 and increases downward to 7.0. Cambisols are widespread in all of Africa, but with different properties, depending on the parent material, the climatic conditions, and the relief. Slightly acidic soils with clay-enriched subsoil and high nutrient holding capacity due to kaolinite, illite, and montmorillonite, good water holding capacity, and a developed soil structure are referred to as Luvisols. They are the characteristic soil type on relatively young surfaces of the subtropical Mediterranean Africa as well as the Cape region of South Africa, and cover four percent of the area of the continent.

In the winter-wet subtropics of the Maghreb and the Cape region are **Chromic Cambisols** and **Chromic Luvisols** widespread, shallow, base-rich, humus-poor, leached soils, which quickly harden during the summer dry season. The Luvisols (partly "Terra fusca") are older, occasionally relict soils, which preferentially occur in the wetter areas on carbonate or silicate mixed rocks. On a small scale, a wide range of distinct soil types can be identified, often characterized by reduced phosphorus and nitrogen contents due to their development on nutrient-poor rocks. These soils are fertile and suitable for arable farming, but their use is limited by the summer dryness, which necessitates irrigation. The clay-rich Cambisols (partly "Terra rossa") developed from carbonate rocks. With irrigation, year-round arable or horticultural use is possible. However, in the Maghreb, the degradation of the potential natural vegetation due to the frequent heavy rains in the winter half-year leads to a high vulnerability to erosion. Since deforestation was carried out ubiquitously in the mountains very early in antiquity and even more intensively since the Middle Ages, the soils in large parts are eroded, their profiles truncated, and often already completely eroded. The soil cover in higher elevations is widely characterized by degraded A-horizons on the one hand and colluvium in depression areas on the other. Cambisols often occur in combination with Vertisols (Tirs).

With annual precipitation of less than 500 mm and less than three humid months of the dry tropics and subtropics of Africa, the soil inventory is limited to sandy, easily erodible soils with low water and nutrient holding capacity and moderate horizon formation. They are referred to as **Arenosols** and cover 22 per cent of Africa's area. Arenosols are the most widespread soil type in Africa. They form in situ through the weathering of quartz-rich parent material or on previously deposited sands (dunes, sand pits, deltas). Due to the low chemical weathering, however, soil formation is limited. Without vegetation cover, Arenosols are extremely susceptible to wind erosion. If vegetation has formed, however, the accumulation of organic materials, clay bands, or the formation of humus-aluminum complexes is possible. Then, with irrigation, the cultivation of peanuts, alfalfa, vegetables, wine, melons, or even corn is possible. Rainfed agriculture, on the other hand, is only possible with annual precipitation from at least 400–500 mm and additional humus and mineral fertilization. **Calcisols** form after the supersaturation of the soil solution or by evaporation due to the leaching of carbonates from the topsoil (solum) and the subsequent precipitation in the subsoil (substratum). Soils with significant accumulations of secondary carbonate are formed. This $CaCO_3$ fills the pores up to the formation of a so-called "hardpan", which is impenetrable for plant roots. Calcisols occur mainly in the dry areas of North Africa, East Africa, and Southern Africa (Namibia, Botswana, Zimbabwe). Gypsisols (on two percent of Africa's area) are characterized by a significant accumulation of secondary gypsum. They are found in the driest areas of the arid regions, including in the northern Sahara, especially in depressions or former Holocene Lake areas and also in the coastal Namib, where gypsum was blown out from the glacially dried-up shelf (Kempf 2000).

In the southern Sahara and the Sahel, one encounters reddish-brown and light brown, often calcified soils. Even further south, in the wet-dry zone of the Savanna, on red soils. These are **palaeosols**, which did not form under the current climatic conditions but have ubiquitously developed in times of higher precipitation, last in the Late or End Pleistocene and the early Holocene, i.e., soils that have already undergone a weathering and soil formation process. Hematite and other water-poor, red-colored iron oxides are very stable compounds, whose reddish soil color has been preserved to this day. In the current wet-dry tropical and dry subtropical zones of Africa, one often sees soils from old weathering residues, which indicate moister (palaeo-) climatic conditions, and recent soils from unaltered outcropping rocks of a light brown color. These soils, so different in structure and composition, can only be understood from the Climate history. Thus, Felix-Henningsen et al. (2009) were able to assign the paleosols occurring in the southern central Sahara and the Sahel of Niger to three different moist periods with soil formation in the Late Pleistocene and Holocene. During the Holocene, weathering intensity and the thickness of the soils—Cambic Arenosols in the Sahara and Luvic Arenosols in the Sahel—decreased from southwest to northeast. The remnants of the paleosols thus indicate, as climate indicators in the early and middle Holocene, a zonal pattern of humidity, with a pronounced gradient from west to east and from south to north.

**Solonchaks** are found on about one percent of Africa's surface area, only in hyperarid regions of the dry subtropical zones and in areas where saline groundwater occurs near the surface (Fig. 7.3). These are soils with high contents of easily soluble salts. Soils with accumulations of clay minerals and exchangeable sodium in the substratum are referred to as Solonetz. They are highly alkaline and have a dense, clay-rich substratum with high exchangeable sodium portions. Clay relocation processes in combination with high pH values are typical for these soils in the lowland areas of the Sahara, in the northern parts of the Sahel, and southwestern Africa, and in salt-rich coastal deposits.

In the moist subtropics of southern Africa, Acrisols and Alisols are common. The former are characterized by leaching and clay accumulation in the subsoil (argic B-horizon). They also show a lower base saturation and base exchange capacity than Luvisols, a deep weathering with base leaching and, associated with this, a strong acidification. Therefore, Acrisols can only be used permanently for agriculture with continuous fertilization. Alisols, on the other hand, have higher proportions of chlorites, smectites and vermiculites, and thus have a higher base exchange capacity than Acrisols. Alisols also have a low base saturation and a correspondingly extremely acidic soil reaction.

**Azonal soils** cover about ten percent of Africa's surface area. They are less developed, and their genesis is significantly influenced by the soil-forming factors, such as parent material (lithology), relief or soil water balance. The most widespread are Vertisols, deep, clay-rich, mostly dark soils with pronounced swell-shrink dynamics (Hydroturbation/Peloturbation), which leads to a deep-reaching reshuffling and homogenization of the soil material (self-mulching effect). Necessary for their formation are alternating wet and dry phases (see

**Fig. 7.3** Sebhka of Seggedim, Central Sahara, Rep. of Niger. *Photo* R. Baumhauer

Fig. 7.4). Following this, Vertisols in Africa are mainly found in the periodically wet regions. They are mostly developed from marly clays or fine-grained basaltic weathering products in plateau landscapes, valley depressions, and sink areas, such as Bas Fonds or Dambos. In North Africa, Vertisols are referred to as Tirs, and in East Africa as Mbuga. On one hand, they are difficult to work with, as it depends on the exact timing of plowing and sowing due to the alternation of swelling and shrinking. They swell upon water absorption because the clay minerals, mostly smectites (often montmorillonite), incorporate water molecules into their layer lattice structure. In the dry season, the solum shrinks from the top down because the clay minerals release the water upon drying. Therefore, in agriculture, Vertisols are also referred to as "minute soils". They are problematic for transportation during the rainy season because they are hardly passable when saturated—they are then also referred to as "breakthrough". On the other hand, organic material accumulates in the shrinkage cracks and incorporates into the soil structure due to the swelling and shrinking processes ("**surface mulching**" or **Peloturbation**). The swelling-shrinking dynamics often lead to the formation of a characteristic **Gilgai microrelief** of hummocks and hollows (Fig. 7.4). If mechanized cultivation is possible, Vertisols show a high yield, for example in the cultivation of cotton as a "cash crop" in Sudan and in the Highlands of Ethiopia, but also of cereals (teff), corn or rice.

**Leptosols**, shallow soils from bedrock or very skeletal soils (gravel, slope debris) with poorly developed structure, occur on exposed bedrock or debris material, especially in mountainous terrain and in deserts. Regosols represent barely developed soils from clayey-silty loose rock

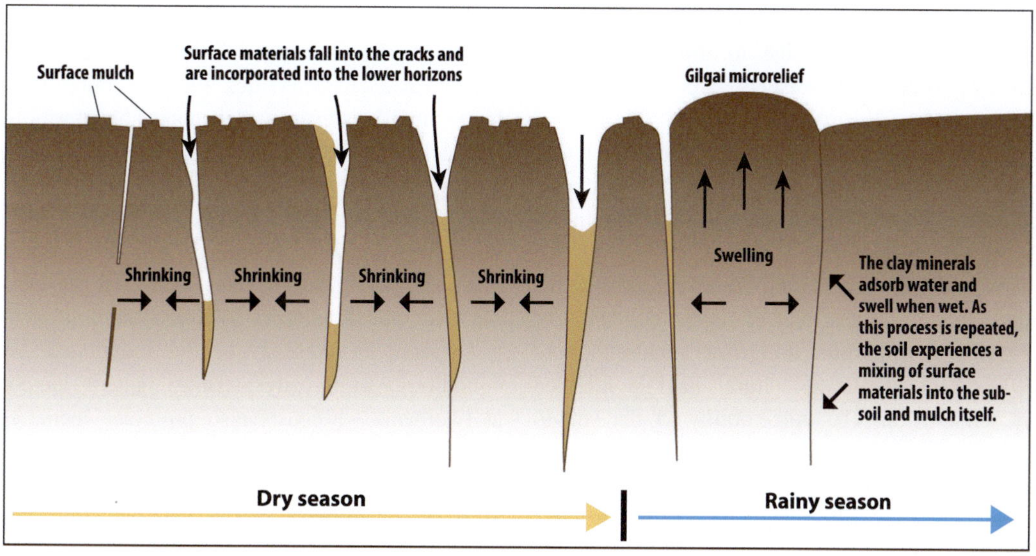


**Fig. 7.4** The soil-forming process of hydroturbation (also referred to as peloturbation) of Vertisols. Modified from Bridges (1979)

with weak profile differentiation, which are also found in mountain regions and deserts. Fluvisols, on the other hand, are young azonal soils from stratified alluvial deposits, which are ubiquitous in Africa along riparian strips and coasts. In contrast to the soils influenced by groundwater either seasonally or year-round (Gleysols) with their typical oxidation and reduction horizons, the floodplain and marsh soils are agriculturally usable for rice and vegetable cultivation as well as for agroforestry. Since Africa has only a few lowland rivers or deltas, Fluvisols are particularly found in the isolated inland basin landscapes, for example in the Okavango Delta, in the Niger Inland Delta or the flood areas of the Barotseland on the Zambezi. They are currently hardly used for agriculture. Andosols are developed in the Rift Valley in East Africa as well as around the area of Mount Cameroon, and in Madagascar on young volcanic ashes. These young soils are generally favorable for agricultural use but are endangered by wind erosion during dry periods. In addition, a partial aluminum toxicity due to mobilization of Al-ions and the associated displacement of essential plant nutrients reduces yield.

## 7.2 Soil Degradation and Soil Erosion

Soil erosion, the most extreme form of soil degradation in Africa, is a major impediment to land use and development. It involves the removal of topsoil or upper horizons by water or wind and is one of the biggest obstacles to land use and development on the continent. Sparse or completely missing vegetation cover and frequent heavy rainfall events as well as human overuse, for example, through overgrazing in dry areas due to strong population growth and/or the expansion of grazing areas after a series of years with high rainfall followed by "normal years", often also accompanied by droughts, can lead to complete vegetation destruction. In their wake, dunes that were once fixed by the vegetation cover are mobilized, and in the worst case, as has happened repeatedly in the Sahel in recent decades, even intact garden and grazing areas are destroyed by the mobilized dune sands. In addition, extensive **shifting cultivation** in the tropical rain forest and the Savanna woodlands accelerates, in addition to nutrient deficit, the devastating water erosion. Ultimately, due to inappropriate land use, i.e., overuse, in Africa, in most cases, often due to rapid population growth, "**man-made deserts**" are created. Even landscapes with sufficient rainfall for rainfed agriculture become desolate due to deforestation and soil erosion, and "dry out". This is exemplified by large parts of the Atlas Mountains in Morocco and Algeria.

Worldwide, about ten million hectares of arable land and 75 billion tons of fertile soil are lost annually due to **water and wind erosion**. About 90 per cent of all agricultural land is affected by soil erosion. The vulnerability of the soil to erosion depends on a combination of several factors. These include soil structure, seasonal vegetation cover, topography, geology, hydrology, climate, and several other factors. These not only influence the vulnerability to soil erosion but also the erosion rate. Soils with a medium to fine texture, a low content of organisms, and poorly developed, non-binding structures are most vulnerable to erosion. The more vegetation cover is present, and thus water can be absorbed, and the topsoil (solum) is consolidated by their roots, the better the soil is protected from erosion. The interception effect against the splash effect, i.e., the substrate mobilizing raindrop impact energy, is higher, the denser the coverage with vegetation, but also litter. About 60 per cent forested area is needed to protect the soil from extreme erosion, but also landslides. In addition to vegetation, relief also plays a significant role: The sensitivity to water and wind erosion and the actual erosion rates are particularly high in marginal areas and steep-sloped terrain. Thus, the current losses due to soil erosion around the steeper slopes of the Central African Ridge in Rwanda, Burundi, and the Kivu province of the Democratic Republic of Congo without soil-protecting measures, are about 100 tons per hectare annually. In the flat relief of the humid Savannas of West Africa, the soil erosion under the natural potential Savanna vegetation is only about 0.9 tons per hectare per year. If the vegetation is cleared, this erosion value increases to 14 tons per hectare. In Central Europe, according to Schröder et al. (2002), the annual erosion over loess is, on average, up to four tons per hectare. Geological conditions, such as the resistance of the parent material, influence the strength of the erosion, as do hydrological characteristics. Frequent rainfall and high rainfall intensities also make the prevailing climate a central influencing factor. If more water accumulates, the erosion rate increases with corresponding runoff. However, with increasing rainfall, the rate of soil development and vegetation density also increases. It turns out that well-developed soils are more resistant to erosion. In addition to the mentioned site conditions, an already existing natural erosion at riverbanks or coasts also promotes increased erosion. In addition, agricultural methods and irrigation, as well as other human activities such as road construction, active soil extraction, and other construction measures. Since the erosion rates are extremely variable in their expression, a generalization of soil erosion and the degradation status for the entire Africa is not possible. Many African governments are now aware of the problem and have created erosion hazard maps and associated models, such as the Soil Loss Estimation Model for Southern Africa (SLEMSA), which was developed in Zimbabwe and adapted in many countries (Salumbo 2020) (Fig. 7.5).

Soil erosion can be distinguished between **On-Site and Off-Site effects**. On-Site effects occur and act through damage directly on the affected areas. Off-site effects, on the other hand, cause damage to areas outside of the affected ones. While erosion rates are difficult to measure, their effects can be well characterized: Changes in soil quality, nutrient losses, crop declines, loss of biodiversity and seed banks, financial impacts, national economic costs, as well as social impacts are analyzed. In addition to these consequences for humans, those for nature and the ecosystem are crucial. Soil erosion increases the surface runoff and reduces the infiltration rate and the water storage in the soil. Organic components and vital nutrients are removed from the soil, and the soil thickness shrinks, which in turn reduces plant growth and biodiversity and simultaneously triggers a self-reinforcing effect. This leads to the reduction of all important soil functions, especially soil fertility and buffer or storage capacity, up to the

**Fig. 7.5** Soil erosion nearby Loskop in the Drakensberg area, KwaZulu-Natal, Republic of South Africa. *Photo* E. Nathan, Mauritius images

destruction of the entire soil substance. In addition, soil stability is reduced by shrinking the available root space for plants.

Concerning the yields, the decrease in soil fertility is particularly evident in the tropics as one of the most significant restrictions. The nutrient loss is most intense in East Africa, followed by the West African coastal areas and Southern Africa. The loss is least in the Sahel and Central Africa. In addition to this loss, there are also general consequences for entire ecosystems in the form of extensive sedimentation in reservoirs or, for example, the increase in suspended load in rivers. Added to this is the not-to-be-underestimated contribution of climate change, especially rising temperatures. The oxidation of high rates of organic carbon, which was previously bound in clay and organic material in the soil, leads to the release of additional $CO_2$ into the atmosphere, which leads to warming and, thus, indirectly to the intensification of precipitation. These, in turn, lead to further increased erosion and consequently to even more carbon release.

Even though regional and global statements about the areas affected or at least influenced by soil erosion are very heuristic and, due to the derivation from small-scale studies, always to be considered as approximations, some basic assumptions and impact chains can be derived. Thus, erosion rates are often ten to twenty times higher than soil formation rates. Depending on the source, the global erosion values on agricultural areas are between 24 and 75 billion tons per year. Since the geographical distribution and size of the areas affected by soil degradation and soil erosion are only approximately known, the international project GLASOD ("Global Assessment of Soil Degradation") was launched, and a global map of the distribution of anthropogenically caused soil degradation was developed from 1988 to 1991. Water erosion represents, in its consequences, the most catastrophic form of erosion. Studies show that a few heavy rainfall events are responsible for the majority of the annual erosion rates. In Zimbabwe, field investigations showed that in extreme cases, up to 75 per cent of the erosion within ten minutes, and 20–40 per cent of the topsoil was removed during a single event. In particular, the **areal degradation** (**sheet erosion**) is important as it directly and indirectly affects soil functions and leads to the formation of erosion grooves, washouts, and erosion channels. Such gully structures are also called "Dongas" in Africa and can reach depths of two to six meters, and even more in mountainous countries like Lesotho. In southern Africa, sheet erosion is one of the biggest environmental problems. The gullies mainly form in the catchment areas of the upper courses of large river systems, often also on the middle courses of smaller tributaries, and are very often the most important source of sediment there. Gullies contribute 50–80 per cent to the total sediment production and represent a preliminary stage to the formation of "Badlands", i.e., a terrain completely cut up by erosion channels or rills, named after the area of the same name in South Dakota, USA. **Gully erosion** is often caused or increased by inappropriate land use combined with heavy rainfall events, which lead to increased surface runoff and intensified sheet flushing processes, often with erosion rates of several decimeters within a few years. If Keystone Species are affected by soil erosion, entire ecosystems can change or be destroyed completely in a short time. As a result, the frequency of species with a special functional and definitional importance for their respective community decreases or they disappear completely, allowing "**degraded**" **ecosystems** to establish themselves.

Erosion processes by wind primarily occur in arid and semi-arid areas with little to no or seasonally very fluctuating vegetation cover. Wind erosion acts over a wide area and leads to a permanent removal of the topsoil (solum), the organically shaped layer, up to the exhumation of the substratum (subsoil or B-horizon) with subsequent sanding, desert pavement, hardening, and encrustation of the soil surface. While water erosion is the dominant form of erosion in Africa as a whole, wind erosion dominates in the Sahel zone and the dry areas of southern Africa (Fig. 7.6).

In the last 40 years, about 30 per cent of the cultivated areas in Africa have become unproductive. This means that approximately ten million hectares of arable land are lost annually due to soil erosion, making it the main threat to local and regional food production in African countries. In addition to natural soil erosion by water and wind, salinization caused by inappropriate irrigation strategies or **tillage erosion** also results in increased erosion rates. In addition, the highly mechanized agriculture caused by "**land grabbing**" in recent years also leads to an increase. The non-governmental organization GRAIN (Genetic Resources Action International) characterized the term Land Grabbing in 2008, referring to the acquisition of significant agricultural land by international agricultural corporations, private investors, and state actors through long-term lease or purchase contracts for the production of globally marketable cash crops (agricultural crops for sale or export purposes) in parts of Africa, especially in Ethiopia, Sudan, the Democratic Republic of Congo, but also in Tanzania, Mozambique, Liberia, and Sierra Leone. However, in view of sustainable food security, soil protection and soil conservation are becoming increasingly important. The preservation, but also the replanting of vegetation, the leaving of crop residues as organic cover and a supply of nutrients, as well as mulching and **shifting cultivation** can contribute to an increase and preservation of a dense soil cover as possible. The latter is a spatially extensive, traditional form of agriculture, where fields are intensively used for only a few years and then a relocation of the cultivation areas with **conservation tillage** takes place. Fundamental changes in the topography, such as terracing to reduce the slope inclination, are also suitable as special soil conservation measures. These include, for example, erosion and washout inhibiting hedges, the introduction of **strip cropping**, the creation of contour walls and parcels of forest land, as well as strips of coppice on watersheds or crests. Currently, a

**Fig. 7.6** Unsuccessful attempt of dune stabilization on the southern edge of the Central Saharan oasis Bilma, Rep. of Niger. *Photo* R. Baumhauer

number of soil protection and soil improvement programs are being carried out. They show very positive effects on yields, water resources, and nutrient availability in terms of sustainability. Also, the advancing climate change affects the change in precipitation and wind conditions, as well as its influence on vegetation cover and biological soil processes in many regions, increasingly accelerating erosion.

**Excursus: Soil Erosion in Madagascar**
Madagascar's extremely high soil erosion rates result from the extensive deforestation (about 90 per cent), in the wake of which many regions degrade, biodiversity significantly reduces, and there are significant deposits in rivers and Lake Alaotra. The latter is now largely filled with sediment and already in 2000 had only 20 per cent of the original area. Throughout Madagascar, there is widespread erosion of the slopes and the formation of gullies (here: Lavakas); in some cases, there are more than 30 gullies per square kilometer with up to 300 m in length and 70 m in width. Even though these occupy only a fraction (<4 per cent) of the land surface, 84 per cent of the river sediments are due to gully erosion and only 16 per cent to hillside denudation. Undoubtedly caused by human activities are about 25 per cent of the examined gullies, at least 20 percent are older than human settlement in Madagascar and both due to tectonic processes and the increasing climatic aridification. However, the numerous anthropogenic interventions also accelerate erosion. Especially on a local level, the effects of soil erosion are of utmost importance, as they often significantly reduce agricultural productivity or cause catastrophic floods on the sediment deposition areas.

## 7.3 Desertification

Desertification, together with its economic and social impacts and side effects, is a major problem in large parts of the world's drylands (see Chap. 5). Primarily due to climatic reasons, these regions are susceptible to the processes of desertification. Nevertheless, there are some generally important elements:

- The hallmark of desertification is the degradation of soils and vegetation, as well as a negative effect on water resources, ultimately leading to desert-like conditions in regions where, due to the planetary atmospheric circulation patterns, deserts would not normally occur.
- Desertification affects drylands with only limited carrying capacity and regeneration potential due to physical deficits, i.e., shortage of water.
- Humans, through inappropriate land use, have a causal relationship with desertification.
- Degradation will cause a decrease in agricultural yields, with famines as possible consequences.

Due to the lack of a universally accepted definition, the spatial extent of desertification is hard to grasp. The "World Map of Desertification" (1977) published by UNCOD (United Nations Conference on Desertification) is based on the definition reached by the Nairobi Conference of the same year. It describes desertification as the reduction or destruction of the biological potential of regions, eventually leading to desert-like conditions. Three levels of desertification hazards are distinguished, with the priori exclusion of the hyper-arid natural deserts.

Indicators of desertification are changes in the aspect of a landscape due to human influence, either indicating where certain processes are just about to operate or where they have already done their work. With their help, a given area's desertification can be determined. The physical indicators can be divided into four groups:

- **Vegetative indicators**, such as the spotty or areal destruction of plant cover, changing species composition or reduced growth rates.
- **Hydrological indicators**: decreasing soil moisture, sinking ground-water level, lowered ground-water recharge rates.
- **Pedological indicators**: physical and chemical soil changes because of "aridification", formation of duricrusts and other structural and textural changes.
- **Morphological indicators**: among the fluvial indicators like truncated upper slope soil profiles, deposition of soil sediments in talwegs or buried soil profiles of lower slopes, an aeolian indicator would be the reactivation of formerly vegetation-covered sand dunes.

Furthermore, several socio-economic indicators, including decreasing crop yields and amounts of domesticated animals, wood and fodder shortages, changes in settlement and population structure, spatial mobility, social behavior, or increasing political unrest are also affected. However, establishing a causal relationship between these indicators and desertification processes can be challenging, and correlations are often difficult to determine. Desertification comprises complex interactions between biological, physical, political, social, cultural, and economic factors. It may have positive or negative interdependencies, and there may be shifts in each case among the dominant triggers and their consequences, depending on causal and regional differences. Although desertification manifests itself in the landscape as degradation and reduced carrying capacity, the causes will almost always be found in the socioeconomic realm. Triggers or causes of inappropriate land-use leading to desertification processes, especially when coinciding with climatic extremes, may be historical, political, social or economic, such as rapid population growth, unfavorable land leasing terms (e.g. short leasing periods not enticing peasants to sustainable use), insufficient administrative regulation of use, market-orientation instead of subsistence economy or no access of the poor to land with a relevant carrying capacity. The desertification process is started and significantly steered by humans.

Inappropriate land use, i.e., overuse of fields or pastures, deforestation, overexploitation of groundwater reserves, and inadequate irrigation techniques leading to salinization and/or desiccation, will destroy the plant cover, degrade the soils, and reduce the availability and quality of water. The condition of soil and vegetation and its various feedback mechanisms are particularly significant for the onset of desertification. Soil degradation, erosion and induration have negative impacts on the soil water budget, leading to reduced vegetation cover. As a result, the exposed and unprotected soil surface becomes increasingly vulnerable to erosion, accelerating the degradation processes. Damage to the vegetation cover over large areas will cause the aridification of the near-surface climate. This again will enhance soil desiccation and induration by simultaneously reducing infiltration capacity and increasing surface runoff. Accelerated soil erosion mainly affects the solum, i.e., the upper fine-grained parts of the soil profile where humus and nutrients are concentrated. This leads to increasingly unfavorable conditions for vegetation growth. In drylands, if no measures are taken, the outcome will be a desert-like environment, man-induced by desertification and facilitated by climatic conditions. The people in such a region will lose their livelihood and will be threatened by famine, increasing poverty, and thus a growing potential for crises and conflicts. It is generally a fact that desertification processes are a set of highly complex cause-and-effect interactions, different in each special case and each region with respect to the factors and mechanisms, and thus not to be explained monocausally (Fig. 7.7).

In Africa, the regions at the northern and southern limits of the Sahara, in East and Northeast Africa particularly Northern Kenya, Ethiopia and the Somali Peninsula as well as in southern Africa the border areas of the Kalahari and Namib are most affected by desertification or at least strongly threatened. These landscapes cover an area of about 5.3 million square kilometers and thus around 30 per cent of Africa's area

**Fig. 7.7** Overgrazed and mobilized dunes and Acacia-Balanites Savanna near Tabalak in the Sahel of the Rep. of Niger. *Photo* E. Schulz

outside the natural deserts, with about 65 per cent of the total population of Africa.

Due to these physiographic factors and the increasing human pressure on natural resources for survival, the Sahel region of Africa, with the States Senegal, Mali, Burkina Faso, Niger, Chad, and the Republic of Sudan is the region on earth where the public perceives the land degradation and desertification problem most intensely (Fig. 7.8).

The Sahel is already beyond the critical limit of the growing period of rainfed agriculture of 500 mm/year. For agricultural use, the spatial and temporal distribution of rainfall is much more important than the average amount. Together with the high intensity (for example, in July 1963 in Niamey, about 200 mm of rain fell in one hour!), the heavy tropical rains promote intense surface runoff. They are highly erosive, provide little water to plants, and are, therefore, not very effective for agricultural use. With decreasing amounts of annual rainfall, its spatial and temporal variability also increases. The extremely complex interactions often lead to droughts in the Sahel, the effects of which can be catastrophically amplified by human influence.

Historical reports and botanical literature from the eighteenth century to the middle of the last century characterize the Sahel as a more humid Savanna landscape. For example, Lenz (1892) reports on corn cultivation northwest of Sokolo in Mali at 15° N, a region where, currently, even millet cultivation is not possible due to deficient rainfall and degraded soils. Among others, Chevalier (1900) reports on a wildlife-rich Savanna with a high proportion of Sudanese flora elements up to at least 16° N (Savanna predominantly diffusely distributed). In pre-colonial times, relatively short periods of use, sometimes overuse in shifting cultivation, in the land rotation system or nomadic livestock farming alternated with long fallow periods. A low population density and social mechanisms have

**Fig. 7.8** Degraded landscape north of Gouré in the Sahel of the Rep. of Niger. *Photo* R. Baumhauer

prevented widespread degradation of the natural potential. For example, permission for marriage was often dependent on additional soil reserves or additional harvest yields.

Social development since the beginning of colonization, spanning just over 150 years (with decisions and omissions that range from the local to the national to the international level), has led to widespread landscape degradation, culminating in large-scale desertification. For the period from 1900 to 1968, the Sahelian ecosystem was able to partially regenerate after drought phases in the decade around 1908–1916 and the 1940s. At the same time, however, the promotion of market- and profit-oriented production of export raw materials such as cotton or peanuts led to a change in the existing economic, political, and social systems, which resulted in many rural regions no longer being viable based on their previous resource use and the subsistence economy typical for the Sahel.

With the transition from colonial to development policy since the political independence of the affected West African states, which was accompanied by a humid period in the 1950s and 1960s, only the means and instruments have changed. The land use, resource, and spatial development policy has often led to areas that, due to their natural potential, were previously only used temporarily or very extensively and were newly developed. Agricultural policy was and continues to be primarily focused on market, export, and rent production, without considering that for most of the Sahel population, subsistence farming, which has never received any support, was and still is essential for survival. Also, since the founding of the Sahel states, general economic policy has consistently focused exclusively on the cities, while rural areas continued to be neglected—and still are.

The inappropriate land use and the extreme droughts of the 1970s and 1980s, which were accompanied by explosive population growth, initiated and accelerated land degradation and increasingly desertification processes. Numerous studies discuss human influence, i.e., population growth, as the main cause of land degradation and see the decrease in rainfall in the last century as intensifying this influence but not as the primary trigger. Possibly, the land degradation processes would have been contained or even stopped without the population increase, i.e., the ecosystem would have returned to the state of the first half of the last century. Satellite image evaluations from the 1970s and 1980s show this partial recovery for some regions. Pollen analytical studies conducted in Chad reveal that there is no discernible southward shift in flora at the Sahara-Sahel border. Instead, a recovery and stabilization of the vegetation in the southern area of the Sahara is observed (see Schulz 1999). The study areas in Chad remained largely unused for nearly two decades due to the civil war. The southeast region of Niger is another area where research in the 1930s has already demonstrated that the Sahara is not expanding southwards. Instead, the Savanna system in this region has remained stable for at least the past 200 years (see Stebbing 1935). In addition to inappropriate land use, it was the drought of the 1970s that dealt the decisive blow to the Savanna system, from which it has not fully recovered to this day, despite the basic regenerative capacity of soil and vegetation.

With the population increase since 1975 from about 17 million to 34 million at the beginning of the 2000s (in Niger, Burkina Faso, and Mali; 2019 together 65.5 million!), without any improvement in agricultural techniques, the agricultural overuse of cultivated areas, overgrazing, deforestation and forest clearing, as well as the exploitation of groundwater reserves have intensified landscape degradation and in many areas already led to the permanent destruction of soils and vegetation. Although already at the UNCOD Conference 1977 (United Nations Conference on Desertification) following one of the most devastating drought disasters in the Sahel region of Africa (1969–1974), goals such as "Combat Desertification" or the "Spread of ecologically adapted productive forms of land use" were propagated, it soon became clear that the processes leading to increased land degradation and ultimately desertification involve highly complex cause-effect correlations that differ from case to case and from region to region in terms of the factors and mechanisms at work. Despite many great efforts, until today, only minor successes have been achieved in combating it. Visible gradual progress is essentially only at the local and regional level with specifically coordinated and respective climatic trend-adapted countermeasures. In contrast to international and national strategies, these measures do not aim at containing the consequences of degradation and desertification but at eliminating their causes, for whose detection in advance comprehensive, regional monitoring data for precise classification are necessary. Overall, not only the policy of the directly affected states plays a central role in both the causation and the possible combat of desertification. Although to create the scope of political action required, it is limited, domestic reforms and also international changes in economic relations are absolutely necessary to contain the processes (Fig. 7.9).

**Fig. 7.9** Firewood transport in the Sahel of West Africa. *Photo* E. Schulz

The significance of climate change for the desertification problem is difficult to estimate. However, it seems with all the current uncertainties and knowledge deficits as a basic tendency to emerge that the desertification processes without effective countermeasures in the context of "global warming" will further intensify and spatially further expand, especially with further population growth. The IPCC, the Intergovernmental Panel on Climate Change (most recently 2022), assumes that with the increase of changes, the adverse effects, also in the form of an acceleration of desertification, come to the fore. In the context of a general temperature increase, heat waves and dry periods will also become more frequent. These play an important role in desertification processes by increasing the vulnerability of soils to degradation in the affected areas on the one hand and impairing the available water resources both quantitatively and qualitatively on the other hand, thus leading to crop failures and a threat to the water supply. For the regions in the Sahel of Africa already threatened by desertification today, a further reduction of the already low soil moisture is expected. In addition, increased evapotranspiration rates affect infiltration and soil moisture and, thus, the amount of plant-available soil water. More frequent and intense heavy rainfall events also accelerate degradation processes, especially soil erosion. Higher evapotranspiration rates with reduced surface runoff impair the water supply not only in quantitative but also in qualitative terms. Together with potentially declining yields in agriculture, this results in negative consequences for socio-economic conditions, which in turn counteract sustainable use and thus promote desertification.

Against the background of climate change and despite the recognizable willingness and political will at the national and supranational level to fundamentally address the problem (for example, through the Great Green Wall project in the Sahel— "green wall"), it is questionable whether desertification can be permanently combated (Table 7.1 and Fig. 7.10).

**Table 7.1** Indicators and causes of desertification

Inappropriate land use ↘ Population pressure ↓ Danger of drought ↙

**Desertification**

↗ ↑ ↑ ↖

| Physical indicators | Regional indicators | National indicators | Global indicators |
|---|---|---|---|
| Rainfall variability | Scarcity of land | No good governance | Poor world trade conditions |
| Poorly developed soils with sparse vegetation, lack of humus | Short leasing periods not enticing peasants to sustainable use | Insufficient administrative regulation of use | Indeptedness |
| Reduced vegetation cover will accelerate the erosion of the soil surface | No access to land with a relevant carrying capacity | National pricing policy is hostile to producers | Low priority for international financial aid |
| | Market orientation instead of subsistence economy | | |

**Fig. 7.10** Reforestation campaign 1986 in the Niger Valley near Niamey, Rep. of Niger. *Photo* E. Schulz

## References

Bridges EM (1979) World soils. Cambridge University Press, Cambridge, p 128

Chevalier A (1900) Les zones et les provinces botaniques de l'Afrique occidentale francaise. C R Acad Sci 130:1205–1208

FAO (2002) World reference base for soil resources. World soil resources report 84. Rom, p 100

FAO-UNESCO (1977) Soil map of the world, Africa, vol VI. UNESCO, Paris

Felix-Henningsen P, Eberhardt E, Kornatz P (2009) Paleo-climatic interpretation of soil development on Sahelian ancient dunes of different age in Niger, Chad and Mauritania. Palaeoecol Afr 29:91–106

IPCC (2022) Summary for policymakers. In: Climate change and land, an IPCC special report on climate change, desertification, land degradation, sustainable land management, food security and green house gas fluxes in terrestrial ecosystems

Kempf J (2000) Klimageomorphologische Studien in Zentral-Namibia: ein Beitrag zur Morpho-, Pedo- und Ökogenese. Dissertation Fak. Geowiss. Uni. Würzburg, p 717

Lenz O (1892) Reise durch Marokko, die Sahara und Sudan. Zweiter Band, Leipzig

Salumbo A (2020) Review of soil erosion estimation methods. Agric Sci 11:667–691

Schröder D, Niemeyer J, Schneider R, Niebes D, Rothstein B (2002) Umweltgerechte Bodennutzung zur Reduzierung von Bodenverdichtung, Hochwassergefahr und Erosion sowie Nähr- und Schadstoffbelastung in einem Mittelgebirgsraum. Ergebnisbericht SFB 522, Universität Trier 121–144

Schulz E (1999) The southern margin of the Sahara in the Republic of Tchad. Vegetation, soil, and present pollenrain. Zbl Geol Paläont 1(5/6):483–496

Stebbing EP (1935) The advance of the desert. Geogr J 91:356–359

## Further Readings

Baumhauer R (2010) Desertifikation: Risikoraum Afrika. In: Glaser R, Kremb K, Drescher A (eds) Afrika. WBG, Darmstadt, pp 47–55

Baumhauer R (2011) Desertifikation und Klimawandel. In: Gebhardt H, Glaser R, Radtke U, Reuber P (eds) Geographie. Physische Geographie und Humangeographie, 2 edn. Spektrum, Heidelberg, pp 1225–1229

Baumhauer R (2018) Einfluss extremer Trockenheit auf Desertifikationsprozesse. In: Lozán JL, Brandt A, Grassl H, Piepenburg D, Wiltshire KH (eds) Warnsignal Klima: Extremereignisse 33–39

Behnke RH, Mortimore M (2016) Introduction: the end of desertification? In: Behnke RH, Mortimore M (eds) The end of desertification. Springer earth system sciences. Springer, Berlin Heidelberg

Breckle S-W (2021) Grüne Mauern ("green walls") gegen die Wüstenausbreitung. Maßnahmen gegen Desertifikation. In: Lozán, Breckle JLS-W, Kasang D, Weisse R (eds) Warnsignal Klima: Boden und Landnutzung, pp 348–355. https://www.klima-warnsignale.uni-hamburg.de

Cox R, Bierman P, Jungers MC, Rakotondrazafy AFM (2009) Erosion rates and sediment sources in Madagascar inferred from 10Be-analysis of Lavaka, slope, and river sediment. J Geol 117(4):363–376

Dlamini P, Orchard C, Jewitt G, Lorentz S, Titshall L, Chaplot V (2011) Controlling factors of sheet erosion under degraded grasslands in the sloping lands of KwaZulu-Natal, South Africa. Agric Water Manag 98:1711–1718

Donovan WG, Casey F (1998) Soil fertility management in sub-Saharan Africa. World Bank Publications, p 480

Frankl A, Poesen J, Haile M, Deckers J, Nyssen J (2013) Quantifying long-term changes in gully networks and volumes in dryland environments: the case of Northern Ethiopia. Geomorphology 21:254–263

Frankl A, Deckers J, Moulaert L, Van Damme A, Haile M, Poesen J, Nyssen J (2014) Integrated solutions for combating gully erosion in areas prone to soil piping: innovations from the drylands of northern Ethiopia. Land Degrad Dev 27(8):1797–1804

Giannini A, Biasutti M, Verstraete MM (2008) A climate model-based review of drought in the Sahel: desertification, the re-greening and climate change. Glob Planet Chang 64:119–128

Giller KE, Witter E, Corbeels M, Tittonell P (2009) Conservation agriculture and smallholder farming in Africa: the heretics' view. Field Crop Res 114:23–34

Gonzales P (2001) Desertification and a shift of forest species in the West African Sahel. Clim Res 17:217–228

Herrmann SM, Hutchinson CF (2006) The scientific basis: linkages between land degradation, drought and desertification. In: Johnson PM, Mayrand K, Paquin M (eds) Governing global desertification: linking environmental degradation, poverty and participation. Ashgate, pp 11–25

Herrmann SM, Hutchinson CF (2005) The changing contexts of the desertification debate. J Arid Environ 63:538–555

Jones A, Breuning-Madsen H, Brossard M, Dampha A, Deckers J, Dewitte O, Gallali T, Hallett S, Jones R, Kilasara M, Le Roux P, Micheli P, Montanarella L, Spaargaren O, Thiombiano L, Van Ranst E, Yemefak M, Zougmoré R (eds) (2013) Soil atlas of Africa. European Commission, Publications Office of the European Union, Luxembourg

Kusserow H (2017) Desertification, resilience, and re-greening in the African Sahel—a matter of the observation period? Earth Syst Dyn 8:1141–1170

Le Roux JJ, Newby TS, Sumner PD (2007) Monitoring soil erosion in South Africa at a regional scale: review and recommendations. S Afr J Sci 103:329–335

Millington AC, Pye K (1994) Biogeographical and geomorphological perspectives on environmental change in drylands. In: Millington AC, Pye K (eds) Environmental change in drylands: biogeographical and geomorphological perspectives. John Wiley & Sons, Chichester, pp 427–441

Morgan RCP (2005) Soil erosion and conservation, 3rd edn. Blackwell, Malden

Mortimore M, Turner B (2005) Does the Sahelian smallholder's management of woodland, farm trees and rangeland support the hypothesis of human-induced desertification? J Arid Environ 63(3):567–595

Nicholson SE, Tucker CJ, Ba MB (1998) Desertification, drought, and surface vegetation: an example from the West African Sahel. Bull Am Meteor Soc 79(5):815–829

Oldeman LR, Hakkeling RTA, Sombroek WG (1991) World map of the status of human-induced soil degradation: an explanatory note, 2nd edn. Global assessment of soil degradation GLASOD. International Soil Reference and Information Centre, United Nations Environment Programme

Pimentel D, Burgess M (2013) Soil erosion threatens food production. Agriculture 3:443–463

Raveloson A, Székely B, Visnovitz F (2017) Gully erosion in Madagascar: causes and impacts. Geophys Res Abs 19. EGU General Assembly 2017

Rockström J, Kaumbutho P, Mwalley J, Nzabi AW, Temesgen M, Mawenya L, Barron J, Mutua J, Damgaard-Larsen S (2009) Conservation farming strategies in East and Southern Africa: yields and rainwater productivity from on-farm action research. Soil Tillage Res 103:23–32

Skowronek A (1987) Böden als Indikator klimagesteuerter Landformung in der zentralen Sahara. Relief, Boden Paläoklima, vol 5, p 184

Smaling EMA, Nandwa SM, Janssen BH (1997) Soil fertility in Africa is at stake. In: Replenishing soil fertility in Africa. SSSA Special Publication no. 51, Madison, pp 47–61

Tamene L, Vlek PLG (2008) Soil erosion studies in Northern Ethiopia. In: Braimoh AK, Vlek PLG (eds) Land use and soil resources. Springer, Dordrecht, pp 73–100

Thomas DSG (1997) Science and the desertification debate. J Arid Environ 37:599–608

Valentin C, Poesen J, Li Y (2005) Gully erosion: impacts, factors and control. CATENA 63:132–153

Warren A (1996) Desertification. In: Adams WM, Goudie AS, Orme AR (eds) The physical geography of Africa. Oxford University Press, New York, pp 342–355

Wessels KJ, Prince SD, Malherbe J, Small J, Frost PE, Vanzyl D (2007) Can human-induced land degradation be distinguished from the effects of rainfall variability? A case study in South Africa. J Arid Environ 68(2):271–297

Zavada MS, Wang Y, Rambolamanana G, Raveloson A, Razanatsoa H (2009) The significance of human induced and natural erosion features (lavakas) on the central highlands of Madagascar. Madag Conserv Dev 120–127

# 8 Biogeography

Africa encompasses a vast spectrum of environments, spanning the entire range of moisture and temperature gradients. From the arid extremes of the Sahara to the lush, well-watered regions, and from the coolness of the Cape to the furnace that is the Sahara, the continent's diverse environments are as varied as they are extreme. In Africa, due to the characteristic latitudinal position, the environmental diversity is mirrored in the proliferation of its flora, for Africa has seemingly eight different biomes, or nine if the anthropogenically influenced forest-poor highlands of the upper Karoo in Lesotho are included. They range from the **equatorial tropical rain forest** to the surrounding **savanna woodland and grassland** to the **tropical and subtropical semi-deserts and deserts**. The northwest and southwest edges of the continent show the characteristic **Mediterranean hard-leaf and maquis vegetation**. In addition to these major vegetational zones, in the highlands and mountain massifs of High Africa, **altitudinal zones** corresponding to the tropical high climates can be differentiated (Fig. 8.1).

## 8.1 Tropical Rain Forest

The natural vegetation of the wet equatorial climate is the **tropical rain forest**, also known as equatorial forest, tropical lowland forest, or tropical evergreen forest. It covers about 50 per cent of the area between 3° N and 5° S and west of 30° E and occupies the main part of the Congo Basin. Its northern and eastern borders are quite sharply and straightly formed, while it is less clearly demarcated to the south. To the west, the rainforest extends on both sides of the equator to the coast of Gabon and continues along the Lower Guinea coast through southern Cameroon to southern Nigeria. Then, in southern Benin and Togo, there is a dry coastal gap. There, due to lower rainfall, deciduous moist forests and savannas extend to the coast (see Chap. 5). To the west of Ghana to Sierra Leone, another closed rainforest area, reaching up to 8° N and over 200 km wide, adjoins the Upper Guinea coast. In eastern Equatorial Africa, evergreen tropical moist forests are found only in smaller areas, mainly in the west of the highlands of Uganda and in mountainous locations with southeastern exposure.



R. Baumhauer, *Physical Geography of Africa*, Geography of the Physical Environment,
https://doi.org/10.1007/978-3-662-71406-5_8

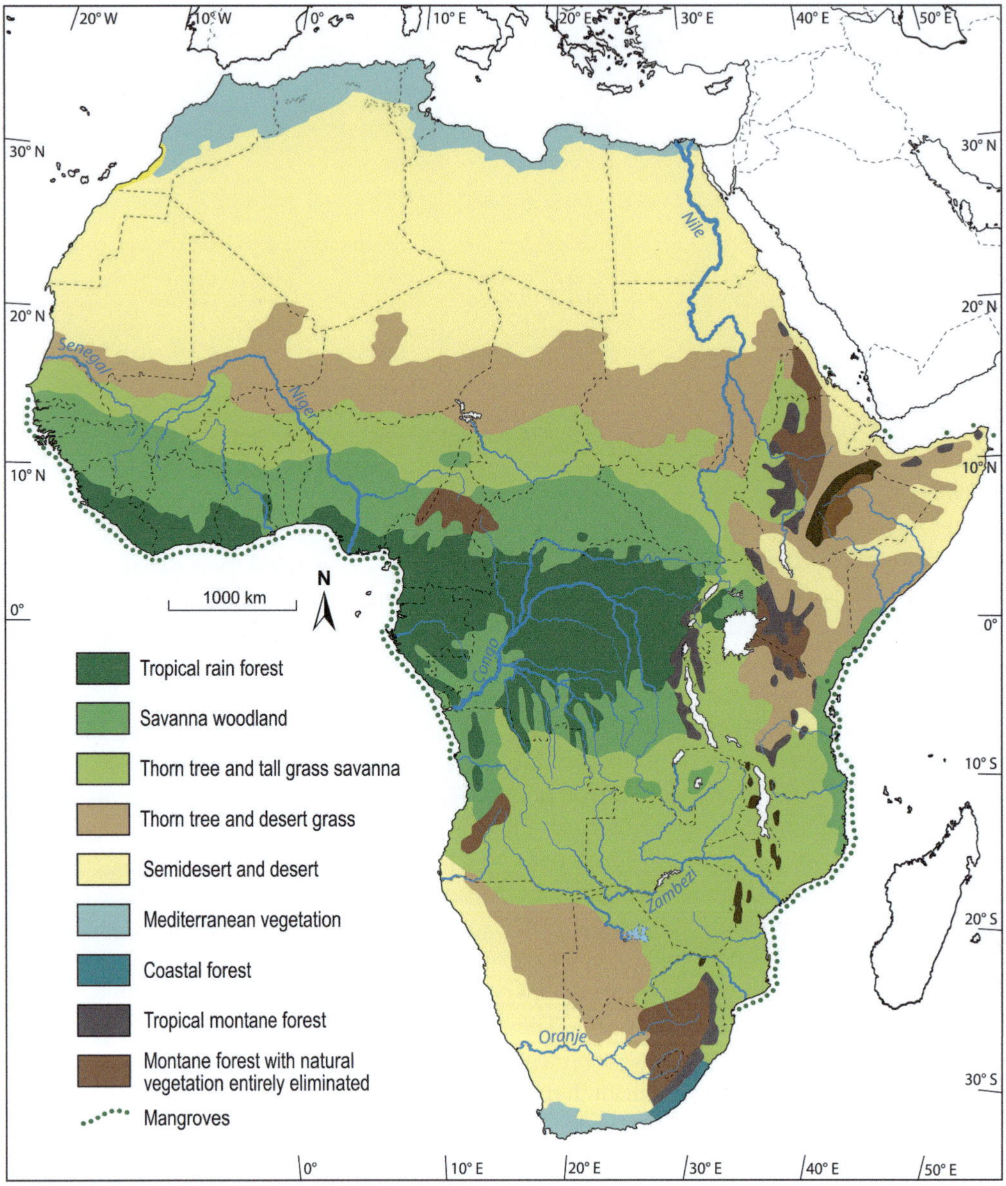


**Fig. 8.1** The natural vegetation of Africa

A characteristic of the tropical rain forest is the high number of trees, which often makes up more than 70 per cent of all life forms. Structurally, the rainforests are complex and multilayered, with the tallest "emergent" stratum consisting of trees up to 60 m in height. A more complete canopy exists in the vicinity of 20–30 m, with various other strata dependent upon the age of the forest and degree of human disturbance. The tree population is very diverse. The species diversity, i.e., the number of species per area, can exceed 100 species per ha (in the forests of the temperate regions of the earth, it is 10–20; each trunk is measured with a diameter of at least 20 cm at breast height). The number of trees per hectare, the so-called stem number, often amounts to several hundred. The trunks are not protected by bark and appear smooth and often light green or white. Many of the largest trees, such as the kapok tree and some Ficus and Annonaceae species (fig and bottle tree plants), are supported at the base of the trunks by plank structures, buttress roots, or prop roots. Apart from trees shrubs, bamboos, and herbaceous elements exist. A characteristic feature of these shrubs is cauliflory, where flowers and fruits grow directly on the stem. Since there are no seasons, the absence of annual rings is characteristic. Lianas do not have to invest energy in the formation of a trunk, as they can grow towards the light with the help of their host plant. They have a growth advantage, especially when they can grow up with the trees. For this reason, these climbers are indicators of the secondary forests. In the undisturbed (primary) forests, however, lianas are rarely found. The epiphyte flora on tree crowns and trunks, as adaptations to the competition factor light, is more diverse than the diversity of life-forms on the forest floor. The epiphytes belong to many different community types. In particular, many epiphytes come from the families of the Bromeliaceae, bromeliad plants, which are also referred to as pineapple plants, and Orchidaceae. Ferns, mosses, algae, and fungi are also well represented in the epiphyte flora. Tree stranglers like figs *(Ficus spp.)* germinate on a tree and completely envelop it, "strangling" it over time. However, only a few of the epiphytes kill their host tree but live with it in parabiosis (see Fig. 8.2).

Since the forest floor only receives about one percent of the sunlight hitting the canopy, hardly any herb layer can develop. As a result of the light attenuation, different air temperatures and a differentiated humidity result for the individual layers. In the area of the canopy, the temperature can be up to 12 °C above the night temperature during the day, and the relative humidity can drop to 30 per cent, leading to considerable, albeit only short-term, saturation deficits. These daily fluctuations decrease rapidly in the lower floors and are hardly noticeable on the ground. The air temperature there both during the day and at night is at the level of the average annual temperature of 25–27 °C with 90–100 per cent humidity (see Fig. 8.3).

Natural tropical rain forests are only found in a part of the potential forest areas in the wet equatorial tropics. In the more densely populated parts, agricultural areas and secondary forests

**Fig. 8.2** Tropical lowland rain forest in Bwindi National Park in Uganda. *Photo* PRILL, Mauritius images

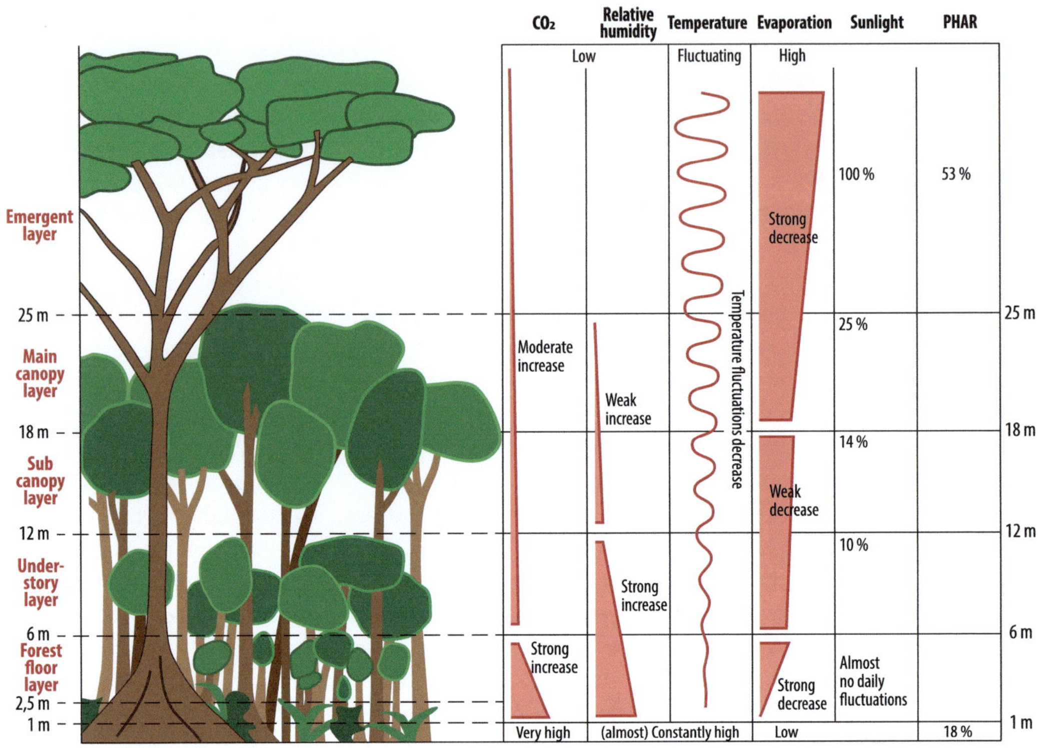


**Fig. 8.3** The typical structure of a low-latitude rain forest (Ivory Coast, Republic of Côte d'Ivoire). *Source* Bourgeron (1983) and Schultz (2002)

often predominate so much that the original extent of the rainforest and the boundaries of its potential distribution are hardly determinable. In southern Nigeria, as well as in the other Guinea countries, only scattered protected relics of the originally contiguous forests remain, and it is feared that the forest destruction will continue in the future due to increasing population pressure and the associated land use pressure. In 2017 alone, 1.47 million ha were cleared in the Democratic Republic of Congo, an area almost the size of Connecticut.

Although in Liberia the rain forest has been pushed back by more than half of its natural area in the past 20 years due to shifting cultivation, slash-and-burn, logging, and the establishment of rubber plantations, it comes closest to the natural ecosystem of the tropical lowland rain forest in the central areas of the Congo Basin due to the relatively low population density. In Sierra Leone, the tropical lowland forest and forest ecosystems have been extensively degraded due to widespread slash-and-burn and shifting-cultivation practices. Only in the highlands in the border area to Liberia have isolated relics been preserved. Optimistically calculated, they cover about two percent of the country's area. In the DR Congo, the challenging accessibility of the region has, until now, helped to limit further environmental degradation.

Often, **secondary forest formations** have taken the place of the primary forests. They show large regional and local differences in their species composition. The secondary forest consists of few, fast-growing species and is significantly less diverse and, therefore, more monotonous than the original rainforest. It has a less pronounced stratum of the crowns but a denser herb layer, and is often more difficult to pass than the original rainforest. Secondary forests arise without human intervention, and the forest forms there by itself. In Benin and Nigeria, secondary forests have been entirely replaced by agricultural land, resulting in widespread deforestation. In addition to the establishment of cash crops for coffee, cocoa, palm oil, coconut, or rubber since colonization, farming produced by small or family businesses has also been increasing recently. Responsible for this is the influx from the more densely populated and, due to the enormous population growth, under considerable land use pressure standing northern Savanna regions. Even attempts at reforestation are far from sufficient and will not be able to halt the ecologically catastrophic process of deforestation, especially since these efforts, should they succeed, will always only lead to secondary forests and can never replace the primary forests.

A concise example of the destruction of the rain forest in the Guinea countries is the Republic of Cote d'Ivoire, whose territory of about 322,500 km$^2$ was originally occupied by a third of tropical lowland rain forest. It extended in the west and east of the country almost to 8° N, in the central part still up to 7° N. In 1956, the lowland rain forest had shrunk to 50,000 km$^2$, and thus to half of its original area. By 1983, only 40,000 km$^2$ remained. Currently, only small remnants are present in the still sparsely populated southwest of the country and the border region to Ghana. They together comprise about 15 per cent of the original area. 85 per cent has been destroyed, with the high deforestation rate not exclusively linked to increasing population growth, and not with the necessity to maintain the livelihoods with small-scale farming. So, it is not just African farmers who are responsible for the drastic loss of rainforest as they strive to produce more food. Upon closer inspection, not all cultivation methods that appear like it at first glance are actually small-scale farming. Rather, nowhere else have palm oil and rubber plantations expanded as rapidly in recent years as in West and Central Africa. It turns out that corporations often employ outgrowers. These are small farmers who produce exclusively for the companies, quasi as employees on their land. In the statistics, this then appears as deforestation due to population pressure but is industrial use. For example, in the Republic of Cote d'Ivoire, the world's leading cocoa-exporting nation, 90 per cent of the cocoa harvest is produced by small or family businesses. Also, in the west and south of Ghana, between the Volta reservoir and the border to the Republic of Côte d'Ivoire, a 300–800 m high hilly landscape, where more than 65 per cent of Ghana's population lives, the export-oriented cocoa production dominates.

For the occurrence of the tropical lowland rain forest, it is less the absolute amount than the distribution of rainfall over the year that is decisive. The rainforest is generally limited to areas with no more than two dry months. However, in regions with exceptionally high rainfall, it can also tolerate a slightly longer dry season, as seen in Liberia. The lowland rain forests have a different structure and composition depending on the regional climatic conditions, the elevation, the morphology, the exposure, the soils, and their soil moisture balance conditions. The different structural formations (e.g. slope forests, narrow forests, riverbank forests, floodplain forests, and swamp forests) can change in a confined space, so their distribution on an overview map is hardly representable. The forests of the steep slopes are generally not as high and often have only two strata. However, they are richer in undergrowth and more intertwined with lianas and, therefore, hard to penetrate. The same applies to the riverbank forests, which are particularly rich in lianas and epiphytic ferns. Floodplain forests and swamp forests usually remain lower than forests of dry soils and are less species-rich. In the middle Congo and its tributaries, floodplain and swamp forests are particularly extensive. However, in the industrial tropical timber industry, only individual valuable hardwoods are cut, which must be transported away via previously laid out clearings and roads. The population, which relies on subsistence farming through extensive shifting cultivation, subsequently expands along these transport routes. Cash crops such as cane sugar, palm oil, cocoa, or rubber, produced industrially on plantations, are still (or again) hardly widespread, mainly due to transportation difficulties or political-economic reasons. Therefore, the central parts of the Congo Basin still consist of largely intact primary rain forests. According to estimates by the FAO, there were about 3.01 million square kilometers of tropical primary rain forest in the Congo Basin in 2011. It is thus the second-largest tropical rainforest on Earth after the Amazon. Of the 1.26 million km$^2$ in the DR Congo, about 1,700 km$^2$ were lost in 2002, and the size of the area disappearing each year has since increased. By 2021, it was already about 5,000 km$^2$. The Ituri National Park in the northeast of the DR Congo, with its approximately 63,000 km$^2$, is one of the last completely intact rainforest flora and fauna areas on earth. About a fifth of the forest is protected by the UNESCO World Heritage Okapi Wildlife Reserve. The easternmost extensions of the Ituri Rain Forest reach Uganda and are protected there in the Semliki National Park. The forest also includes the habitat of the Mbuti Pygmies.

From the eastern Congo Basin, the **lowland rain forests** transition into **montane rain forests** with the rise to the Central African Ridge at a mean elevation of 1,000–1,800 m. In these, tree ferns and the Raphia palm are common. In the peripheral areas of the East African Highlands, the montane rainforest is mainly formed as a narrow forest. At the Ruwenzori Massif and the Virunga Volcanoes, the tropical altitudinal zone follows with the height. The elevations of the Central African Ridge above 1,600 m, especially in the Kivu region and the provinces of Ituri and (Haut-)Uele in the northeast and east of the Democratic Republic of Congo, have been largely transformed into a **(long grass) savanna woodland** due to their favorable climatic location through centuries of slash-and-burn agriculture combined with intensive pasture farming. In Gabon and the Republic of Congo as well, these savannas occupy considerable areas. However, they are not exclusively anthropogenic but are orographically influenced as coastal savanna or geologically conditioned as in the Niari Basin in the south of the Republic of Congo. In Gabon, currently over 700 species of woody plants can be found, which belong to 58 different families. However, in many developed parts of the country, especially on the coast, the species-rich primary forest has been replaced by a much less diverse secondary forest due to the extraction of Okumé softwood and the industrial logging of other precious wood species.

**Subtropical evergreen rain forests** exist in Africa only in narrowly defined mountain gorge areas on the east coast of South Africa. There, the southeast trades cause rainfall for most of the year. However, the biodiversity and height of these forests are lower, and cauliflory is rarer than in the tropical lowlands. Otherwise, the character of the forest is similar.

None of the countries of the Congo Basin and the Guinea coast are completely covered with tropical evergreen forest. In the Central African Congo Basin, the rain forest in the north, south, and southwest, as well as in the West African Guinea coast countries, is surrounded by a roughly 100 km wide band, regions of **tropical evergreen seasonal rain forest**. A characteristic of this border zone of the equatorial rainforest, which still receives 1,400–1,800 mm of annual rainfall, is the periodic change of foliage during the roughly three-month dry season. The vegetation mosaic is characterized by wide extensions of the rainforest as gallery forests along the rivers and lighter shrub formations and grass savannas on the ridges and around the watersheds. The better settlement possibilities compared to the rainforest led to agricultural use in historical times. Due to favorable climatic conditions, the cultivation of characteristic rainforest plants such as cassava and oil palm, as well as cash crops such as corn, sweet potato, and cotton, is possible. Large livestock farming and corresponding pasture farming are due to the spread of the tsetse fly and the deadly Nagana disease not possible. Food production is still almost exclusively traditional **shifting cultivation** in the form of **subsistence farming** with very little market production. In the recent past, the unregulated removal of the valuable timber found there, for example, from the family of mahogany plants (*Meliaceae* spp.), has considerably degraded the remaining remnants. At the beginning of the twentieth century, the teak tree (*Tectona grandis*) was introduced from Asia and has since been grown as plantation wood in the Republic of Côte d'Ivoire, Togo,

Nigeria, Cameroon, Tanzania, and Zimbabwe. The influences of land use on the tropical evergreen seasonal rain forests are most pronounced through slash-and-burn shifting cultivation. As a result, they have been transformed into secondary savannas in all Sudanese countries. Dominant here, too, is still subsistence farming in traditional shifting cultivation in the form of hoe agriculture. Without mechanical assistance, grain and tubers can be grown through rainfed farming, which allows for two harvests, with irrigation, of peanuts and cotton. However, the dry season is too long and the humidity too low for export-oriented tree crops such as cocoa, rubber or palm oil. In the Central African Republic, the Republic of Congo, and Cameroon, coffee of the Robusta and Excelsa varieties is grown in this border zone to the savanna regions. Although, in addition to a few larger plantations, about 25,000 mostly cooperatively organized family businesses in the Central African Republic grow coffee, the country ranks only 40th among coffee-producing countries and supplies less than 0.1 per cent of the world harvest (ICO 2022). In the southern extension of the tropical evergreen forest, secondary forests of shifting cultivation predominate in the DR Congo and northern Angola. Areas where the potential natural vegetation has been preserved due to hunting or for the protection of settlements from fire characterize the tropical evergreen seasonal rainforest as climax vegetation. Similar areas of deciduous and evergreen seasonal rainforests also occur in the coastal areas from Tanzania to the mouth of the Rovuma and on the east coast of South Africa.

**Excursus: The Coastal Vegetation of the Mangroves**

The edaphically conditioned mangrove, which in Africa usually appears as about 10–15 m high evergreen woody vegetation in the tidal area of the flat coasts of the sea, is essentially spread in the zone of the tropical rain forests. But mangroves can also be found beyond this, on both sides of the continent at about 15° N, the southernmost on the east side at over 31° S. The streams of the rainforest area are often accompanied by mangrove vegetation far upstream in their estuary area. Particularly extensive are the mangrove swamps of the Niger delta. On the east side of the continent and the coasts of Madagascar, other species dominate than on the Atlantic coasts. Mangroves such as *Rhizophora racemosa* in West Africa represent an important factor in coastal protection. However, they are being destroyed by logging, by the creation of wet rice fields or more recently by the construction of breeding ponds for aquaculture, for example, for shrimp farms, leading to increased coastal erosion and thus to land loss. One notable exception is the Niger delta, where the tidally influenced areas still support a lush and thriving mangrove ecosystem. It is less threatened by deforestation than by environmental pollution due to the oil production in the delta and offshore (Figs. 8.4 and 8.5).

## 8.2 Tropical Savannas

Savanna vegetation in Africa covers extensive areas between the equatorial forests and the mid-latitude desert to the north and south, and indeed occupies up to 65 per cent of the continent. Savanna prevails in strongly seasonal climates where there is a marked dry season, varying in length from a few weeks to eight or nine months at the end of the gradient. The vegetation consists of a mosaic of community types, the most important diagnostic character being the presence of a continuous stratum of grasses at the ground layer. Trees may vary in abundance from a continuous closed-canopy woodland to more open park-like savannas studded with microphyllous trees, to grasslands with scattered trees or bushclumps, to treeless grasslands.

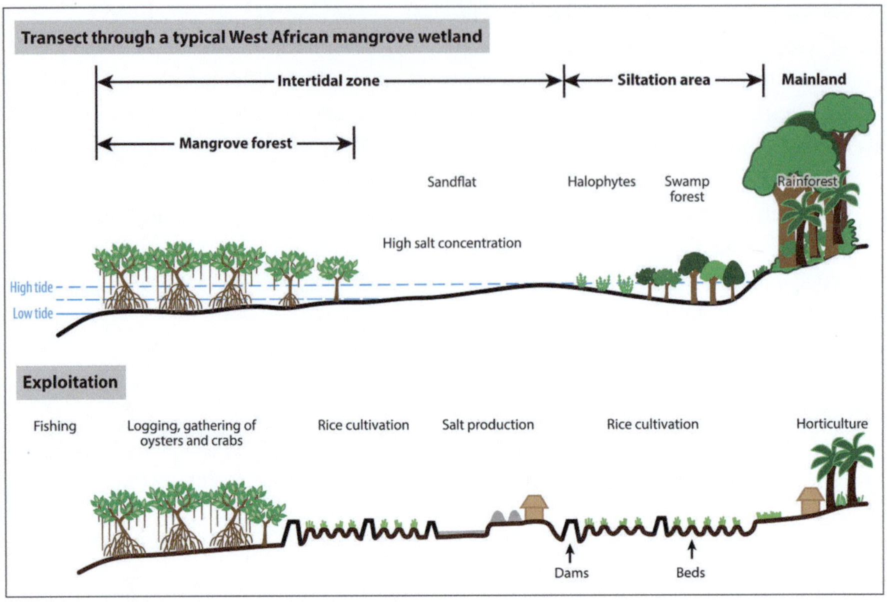


**Fig. 8.4** The coastal vegetation of the mangroves. Modified from Wiese (1997)

**Fig. 8.5** Mangroves in Tanzania. *Foto* Sergejson, Adobe Stock

In West and Central Africa, the Sudan zone with its savanna formations of the **wet savanna (savanna woodland)** and **thorn tree and tall grass savanna (dry savanna or Sudan savanna)** follows in a model zonal arrangement towards the poles, to the north of which in the Sahel zone the **thornbush savanna (thorn tree and desert grass savanna)** follows. However, the individual savanna zones run without clearly visible borders and are, therefore, difficult to separate from each other. As a result of the strong anthropogenic imprinting through livestock farming, agriculture, and the characteristic periodic grass fires set by humans during the dry season, the natural balance of the particularly sensitive savanna ecosystems is almost everywhere in Africa disturbed. South of the equatorial rainforest, the wet savanna belt is only fully developed in the west. Its southern border lies in the interior of Angola at 13° S, on the edge of the Lunda Ridge at 9° S. In northern Angola, it reaches the Atlantic and extends, due to the influence of the cold Benguela Current, beyond the mouth of the Congo River more than 200 km along the coast to the north.

The **wet savanna** forms under natural conditions, i.e., without anthropogenic interventions, with well-developed tree populations and tall grass growth, the wettest of the wet-dry tropics. The northern border of the wet savanna is on the west coast of West Africa at 14° N, in Nigeria at 10° 30′ N, and in Chad south of the Bahr al-Ghazal, in the Central African Republic and South Sudan at about 8° 30′ N. The wet savanna receives 800–1,600 mm of annual rainfall for two to five arid months between May and September. It is the most densely populated region not only of tropical Africa but in West Africa generally. In the extremely densely populated Nigerian Plateau of Jos (formerly known as Bauchi Plateau) and Adamawa, the wet savanna of the Sudan zone, known as the Middle Belt, is largely degraded. The tree population is especially heavily thinned due to firewood extraction and the periodic fire practice.

On vegetation maps, the **wet savannas** are not represented as a unit. Rather, different possible plant formations appear differently in individual subspaces; semi-evergreen and deciduous monsoon forests (tropical deciduous forest or moist monsoon forest), open, rain-green, large-leaved tree shrubs with grass undergrowth (dry monsoon forest or savanna woodland) as well as tree-free tall grass plains (wet savannas) with evergreen gallery forests. The wetland forests of this zone include different forest types, ranging from the semi-evergreen rain forest already characterized above to the predominantly deciduous monsoon forest to the dry monsoon forest. The moist monsoon forest still has evergreen components in the lower strata. The dry monsoon forest, on the other hand, only has single-layered, barely ten-meter-high trees with large-leaved rain-green foliage. It transitions to the open tree shrub of the wet savanna and finally to the treeless tall grass plain. The grass of the open savannas consists of broad-leaved perennial grasses like the Napier grass (*Pennisetum purpureum*), whose leaves in the rainy season form a plant cover up to over three meters high. The wet savanna is also characterized by the presence of gallery forests. These are not only limited to riverbanks and floodplains but also fill valleys and gorges, often up to the upper valley, as they benefit from emerging slope water. They are high forests of the rainforest type with multi-strata canopies and their characteristic tree is the shea butter- or karité tree (*Vitellaria paradoxa*) (Fig. 8.6).

The transition from the **monsoon forests** to the grass-dominated savannas, which are loosely interspersed with shrubs, is already very smooth in the natural vegetation state, and even minor human interventions are sufficient to change the vegetation boundaries. The original natural savanna formations are very much blurred by centuries-old shifting cultivation. The forests of this zone can be cleared much more easily by burning during the dry season than the rainforest. On the other hand, the rainfall is still high enough for a secure harvest. Therefore, the dry forests have largely been converted into anthropogenic savannas ("derived savannas") or permanent farmland. In addition, constant fires have caused severe mineral leaching and soil loss, making it often impossible today to reconstruct the original

**Fig. 8.6** Acacias in the border area of wet and dry savanna in the W-National Park, Rep. of Niger and Benin. *Photo* R. Baumhauer

potential natural vegetation of such savannas. It is certain, however, that the natural savannas, determined by the climate, have occupied a much smaller area than the anthropogenic savannas created under the influence of shifting cultivation and regular burning. Field studies have shown (Menaut 1983) that when anthropogenic land use is eliminated, the number of trees increases in density, height, and species number with the duration of protection. Similar results are available from the south and west of the Democratic Republic of Congo, confirming the **secondary savannafication**. The competitive balance between grasses and shrubs in the savanna is based on the fact that the grasses, with their shallow root system, utilize the adhesive water of the upper soil layers, while the shrubs tap into the occasional seepage water of the deeper soil layers. If the annual precipitation falls below about 300 mm, the holard amounts are too small for the survival of woody plants and the savanna turns into pure grassland. However, if the water balance is shifted in favor of the shrubs by constant grazing of the water-consuming grasses with sufficient water supply, thorny shrubs or thorn bushes gradually take over (see thorn bush savanna) and thereby render the pastureland worthless. In this way, the thorn bush as a replacement community of the climatic savanna has conquered vast areas, such as in the Sahel zone, with only very low carrying capacity as pastureland, which quickly turns into an anthropogenic desert under further stress (firewood extraction, overgrazing). Hornetz et al. (1992) also speak of "Green Desertification" (see Chap. 7 and Fig. 8.7).

In the **dry savanna** (thorn tree and tall grass savanna), the number of humid months decreases to less than six to seven, and the mean annual precipitation is below 1,000 mm. Only two to a maximum of five wet months and mean annual precipitation of 250–500 mm are reached in the **thornbush savanna** (thorn tree and desert grass savanna). South of the equator, the typical

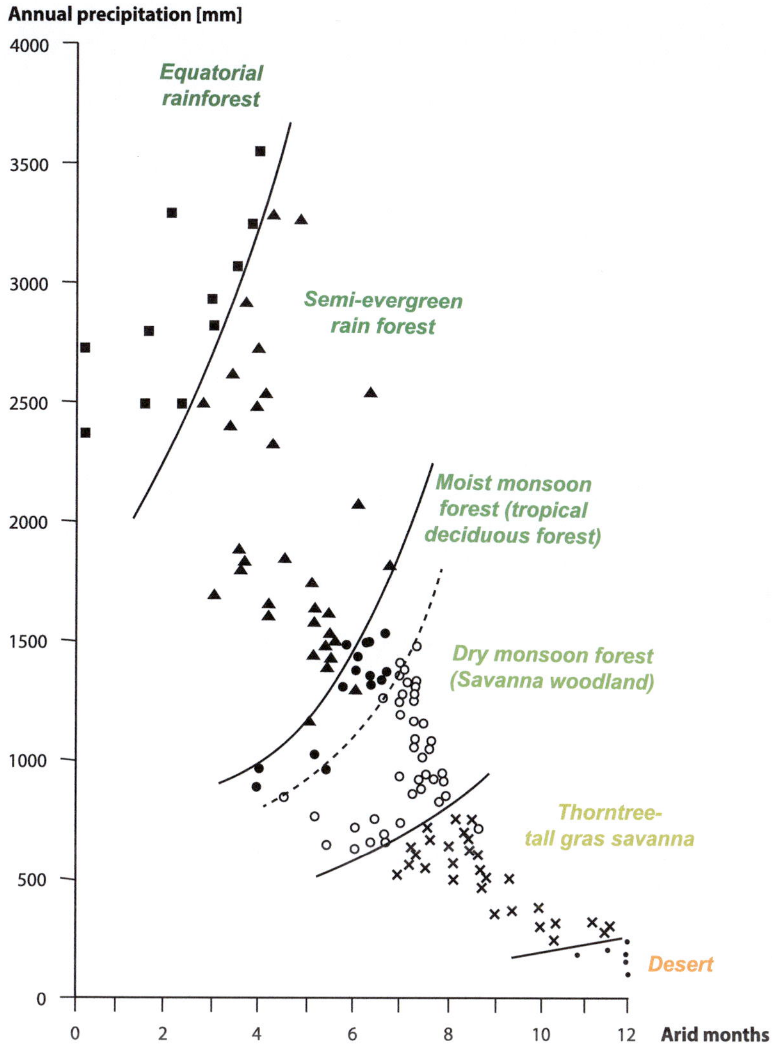


**Fig. 8.7** A comparison of annual rainfall (mm) and annual dry season in months with tropical forest vegetation. Modified from Walther (1990) and Schultz (2002)

latitudinal arrangement of savanna zones, as is characteristic for Africa north of the equator, is no longer present. Here, the climatic contrast of the west and east sides of the continent is crucial for the distribution of vegetation formations. In this vegetation zone, the characteristic distribution of seasonal precipitation is more crucial than the total annual rainfall (Fig. 8.8).

The dry savanna of the northern Sudan zone, naturally with sparse tree cover and low grasses, encompasses a narrow strip around 13° N in the western part of West Africa, in Burkina Faso and southwest Niger from 11–12° N, in Chad around 10° N and around 9° N in the Bahr al-Ghazal of western Sudan. From the West African Guinea countries, the north of Ghana, Togo, Benin, and Nigeria still have a share in this zone. The degree of vegetation degradation has increased significantly since the end of the last century due to the pressure of land use by the growing population in the dry savanna. Although the critical limit of growing period of rainfed agriculture is around 450 mm mean annual rainfall and thus sufficient everywhere in the wet-dry tropics, the mean deviation of the individual annual totals from the long-term mean in the dry savanna can already

**Fig. 8.8** Baobab tree (Monkey Bread Tree, *Adansonia digitata*) in the dry savanna of Mali. *Photo* E. Schulz

be 30 per cent, resulting in highly fluctuating and uncertain yields (see Chap. 5).

Land use continues to be traditional shifting cultivation in the form of hoe farming. Without mechanical or animal assistance, sorghum and other millet species are grown in subsistence farming. Cash crops for the market, such as peanuts, are only planted to a very small extent. To this day, **agropastoralism** prevails, meaning that in addition to crop farming, animals are also kept. A problem in the dry savanna, as well as in the thorn bush savanna, is the increasing pressure of use due to the still unchecked population growth. The often very strong overuse, for example through overgrazing, prevents the transpiration of the grasses and the rainfall penetrating the soil and substrate remains available there until the dry season for the shrubs, and leads to an increasing bush encroachment, which affects large areas, especially of the northern part of the dry savanna. If these shrubs, for example, acacias, are cut down for firewood, almost vegetation-free areas remain. In all Sudan states, the natural potential of the dry savanna is overloaded by the population and pressure of use, and thus, the natural balance, which was also effective as a buffer against drier years, is destroyed. Thus, in South Sudan, there are only very few woodlots left. Here, savannas of semi-high tussock grasses dominate with only sporadic denser, but mostly very sparse tree growth, which consists predominantly of rain-green species with often stately, 12–15 m high thorny umbrella trees, such as the umbrella thorn acacia (*Acacia tortilis*). Only occasionally do succulent trees and palms occur. Savanna fires (see Excursus: Savanna Fires) promote soil depletion. The terrain without vegetation burned off before the rainy season is very susceptible to the wash out of the topsoil and the associated leaching of the soil. The minerals taken from the soil by the plants are leached out of the ash and washed away. Changes in soil structure and texture because of the fires are likely, along with occasional extreme dry years, the reason why no woody plants immediately reestablish themselves, even if the fire is kept away. Studies have shown that after six to ten years without fire, the soils improve, tree seedlings appear, and shrubs of several meters in height grow quite quickly. However, other species settle than the original potential natural savanna shrubs (Fig. 8.9).

In the southern hemisphere of Africa, **Miombo forests** are the typical vegetation

**Fig. 8.9** Heavily overused dry savanna (thorn tree and tall grass savanna) in southern Rep. of Niger with an inselberg in the background. *Photo* R. Baumhauer

formation of the plateau landscapes of southern central Africa. They are ubiquitously preserved and spread across the entire Lunda Ridge of East Angola, North Zimbabwe, and southern DR Congo. In northern Mozambique and central Tanzania, they cover an area of about 1,000 × 1,000 km. They also cover the western edge of Madagascar. The Miombo forest, whose name is derived from the Tanzanian designation of the predominant tree species *Brachystegia* spp., consists largely of large-leaved stands with grass undergrowth. The 10–20 m high trees resemble, in their shape, with straight trunks and broad crowns of the deciduous trees of the temperate zone. In dry areas, the small-leaved umbrella trees with *Berlinia* spp. are more prevalent. Some of these stands are very open, with much shrub growth between the trees. There is also the baobab (*Adansonia digitata*), which is widespread. In the area of the Miombo forests, so-called **Dambos** occur, shrub-free short grass floors, which form in periodically flooded depressions on Petroplinthite (formerly: Laterite or iron pan). In Zambia, Dambos are estimated to cover an estimated 12.5 per cent of the country's area. These flood savannas are edaphic, that is, not tied to a specific climatic vegetation zone. They also occur in the river valleys of the Congo River and the other rivers of Central Africa, and on the upper Nile. In East Africa, they are also referred to as **Mbuga**, in Mashonaland as **Matoro**, in South Africa as **Vlei**, in Nigeria as **Fadama**, and in Sierra Leone as **Bolis**. Regionally, palms (*Hyphaene* spp., for example, Doum palm, and *Borassus* spp., for example, Palmyra palm) are widely scattered on this grassland. The natural riverbanks usually carry fringing forests, often also with palms. East of the Kalahari in South Zimbabwe and Southeast Mozambique is followed by the **Mopane dry forest**, a deciduous forest, consisting of over 90 per cent Mopane (*Colophospermum mopane*), a slow-growing and often multi-stemmed Shrub or small tree up to 12 m in height, which sheds

**Fig. 8.10** Miombo landscape (Brachystegia sp.) and inselbergs in northern Zambia. *Photo* S. Robinson, Mauritius images

its protein-rich leaves, which serve as livestock feed, late in the dry season (Fig. 8.10).

Termites are typical of all savanna formations. The dry forests are interspersed with up to five percent of the soil surface with up to five meters high so-called termite mounds, the nests of the termites. The movement or relocation of soil material by the termites for their buildings ensures a higher humus and plant nutrient content, so this material is gladly used by the hoe farmers for soil improvement.

The **thorn savanna** (also referred to as thorn shrub savanna, thorn bush savanna, thorn tree savanna or thorn bush steppe), includes the areas in the Sahel zone and the Kalahari, in the intra-tropical dry areas of East Africa in Ethiopia, Kenya and northeastern Tanzania and the Somali peninsula and the region in the far south of Madagascar. Common in these areas are high summer temperatures, the absence of effective frosts, and above all, the duration of the dry season with seven to ten arid months and a high rainfall variability in space and in time with annual mean rainfall amounts of 200–700 mm. Depending on the edaphic conditions, this climate can have more or less thorn shrub-rich low xeromorphic grasslands, thorn tree and succulent shrubs, as well as grass-poor shrub formations predominate. The greater the rainfall variability and the stronger the local edaphic conditions increase the dryness, the smaller the proportion of grasslands and the larger the proportion of succulents. The grasslands of the thorn savanna are much lower than those of the dry savannas. The narrow and hard-leaved, usually only knee-high grass tufts stand further apart with widely spaced lower evergreen bunch grasses. The proportion of woody plants is very different. The most common vegetation consists of shrubs, typically one to four meters in height, such as myrrh and frankincense shrubs. Examples are the Egyptian balsam (or Lalob in Sudan, *Balanites aegyptica*), Maerua (*Maerua crassifolia*), Boscia (*Capparidaceae* sp.) or the toothbrush tree (*Salvadora persica*) and individually or in groups growing, usually umbrella-shaped dome trees, which can grow up to 12 m high, like the baobab trees (*Adansonia digitata*) or Acacia communities like *Acacia giraffae*, *A. senegal* (Gum Arabic) or the umbrella acacia (*A. spirocarpa*). The shrubs are predominantly thorny, small-leaved, and rain-green, some also evergreen, like the leafless rod shrubs. Among the grasses, there are sometimes also dwarf—shrubs, succulent semi-shrubs, and in the rainy season, herbaceous plants. A vegetation formation that is only under certain soil and climate conditions in the Sahel and less Common in East Africa is the so-called "**Tiger bush**" ("brousse tigrée"; Clos-Arceduc 1956) with its characteristic vegetation pattern, the alternation of vegetation-free strips with strips of dense vegetation (Fig. 8.11). Tiger bush occurs wherever there is too little rain to ensure

**Fig. 8.11** Tiger bush ("brousse tigrée") near Dosso in southern Rep. of Niger. *Photos* R. Baumhauer

a closed vegetation cover, and slope inclination and surface crusting of the vegetation-free areas lead to surface runoff of the precipitation. The vegetation bands thus receive significantly more water than would otherwise be available under the same precipitation conditions. Following an area without vegetation is a zone with a high proportion of herbaceous vegetation, followed by a center zone with dense shrub vegetation, and finally, a zone where the shrubs die off. In eastern Africa, thorn shrubs and succulent shrubs, as well as grass-poor shrub formations, are found on the Somali peninsula and the western edge of the Ethiopian Highlands, while in the equatorial section of the East African highlands, grasslands dominate. The thorn savanna area, south of the large thorn tree and tall grass savanna (dry forest), mainly includes the Kalahari. Here, grass-poor, very sparse shrub formations occur ubiquitously, both with thorn shrubs and with stem or leaf succulent plants.

**Excursus: Savanna Fires**

Grass fires in the dry season, which occasionally also reach the shrub and tree layer, are characteristic of all savannas. From 2001 to 2009, these fires caused 44 per cent of all carbon dioxide emissions from global fires (Van der Werf et al. 2010). The fires can occur naturally as a result of lightning strikes, but they are more frequently caused by human activity. The fire practice, especially the burning of grass by the savanna inhabitants, has been traditionally carried out in many regions for millennia to maintain and improve pastureland, clear and ash-fertilize cultivation areas, drive away wildlife, and destroy pests. The fire, as a selection factor for flora and fauna, has considerable effects on the ecosystem of the savannas. By altering the heat and water balance of the plant cover and the soil, they influence the energy and material flows, ultimately determining the vegetation structure. Depending on the frequency and timing of the fires in the dry season, various degradation and succession stages are created, which combine to form a vegetation mosaic, a kind of fire climax formation, which only changes when no more savanna fires occur. As a result of frequent savanna fires, fire-resistant tree species (pyrophytes), as well as trees and grasses with a high regeneration capacity, prevail.

The intensity of the fires depends on the amount of combustible material and the weather conditions; that is, the further the dry season has progressed and thus the temperature and humidity have decreased, the more intense these fires can be and thus also affect trees. However, normally only about 70–90 per cent of the grass layer and litter, 20–50 per cent of the litter from trees, 12–60 per cent of the branches and bark, and up to 20 per cent of the dead, but still standing ("standing dead") trees burn (Schultz 2012). Due to the only briefly higher temperatures of the fires, the soil fauna and flora, especially the root system, are only slightly affected. Overall, savanna fires improve the supply of mineral nutrients, which are mainly available to the vegetation in the dry savannas, while they can be leached out in the wet savannas by percolation. Surface and rill erosion often lead to an intensification of erosion on freshly burned areas.

Recent studies show that savanna fires not only reduce the ability of the affected plants to absorb carbon dioxide but also change the soil. They store up to 36 per cent less $CO_2$ than soils that are spared from frequent fires. In addition, the soils contain up to 38 per cent less nitrogen fertilizer after fires. This nutrient deficit reduces the regenerative ability of the seedlings (Deng et al. 2017). However, it should be considered that the actual process depends very much on the further physiographic conditions, depending on the duration and intensity of dry and rainy seasons, extent of animal grazing and other factors, which hinder or promote the regeneration of savanna plants and thus accelerate or slow down the effects of later fires.

## 8.3 Semi-arid and Arid Biomes

Towards the polar margins of the savanna woodland and grasslands of Africa, mean annual precipitation declines below 400 mm, and the length of the dry season becomes a major ecophysiological constraint. It is in these regions that we find the semi-arid and arid biomes of the continent with a more continuous cover of vegetation at the wetter end of the moisture gradient and the true desert at its drier end. Such regions coincide with Sahara, Somalia-Masai, and Karoo. Namib regional centers of endemism and the Sahel and the Kalahari-Highveld regional transition zones. The two major arid regions are the Sahara and Namib deserts where, despite having distinct climatologically causes of aridity, precipitation inputs are sporadic and generally below 100 mm per year (Chap. 5). The terms arid, semi-arid, desert, and semi-desert tend to be rather loosely, but areas in Africa with between 100 and 400 mm per annum precipitation are taken to be semi-arid and include the northern part of the Sahel region to the south of the Sahara and the Karoo and Kalahari of Southern Africa. The former is essentially an open shrubland, and the latter is regarded more as an area of dry thorn savanna (Namib, Kalahari, Karoo, see below The East and South African Highlands).

A contiguous semi-desert belt adjoins with 10–11 arid months and uncertain summer rains of about 50–250 mm per year Sahel and Sahara. In the west, these semi-desert formations continue north along the coast. Semi-deserts and deserts of the tropical-subtropical dry areas cannot always be sharply separated from each other. For edaphic reasons, they can replace each other on a small scale, and the semi-desert often permeates the actual desert with smaller or larger occurrences. The Sahara, the largest desert on earth, is only the westernmost part of the large old-world dry belt, which continues from the Nile countries further over Arabia and Iran to Central Asia. The largest part of the

Sahara receives less than 50 mm of precipitation annually. It falls episodically, with the northern Sahara receiving winter rain from the westerly wind zone, and the southern part receiving monsoonal summer rain. The vegetation is very poor in species with 1,200 species, 25 per cent endemics, and only 450 flowering plant species in the Central Sahara. In addition to the dominant xerophytes, annual floras, which reach flowering and fruit maturity within two weeks after precipitation events, are widespread. **The stone—(Hammada) and gravel desert (Serir, Reg)** are poor in vegetation and show only a sparse plant cover with *Cornulaca* and Saxaul (*Haloxylon ammodendron*). Clay (Dayas) and salt pans (Sebkhas or Chotts) contain halophytes such as glasswort (*Salicornia* sp.), Salsola, and tamarisks. Due to the high groundwater level, the dry valleys (Wadis or Ennedis) are richer in vegetation. There, thorn and rod shrubs appear: Acacias (*Acacia raddiana*, *A. seyal*), *Calotropis procera*, *Leptadenia pyrotechnica*, *Balanites aegyptica*, and *Panicum*. In the **sand deserts (Erg or Edeyen)**, which cover around 20 per cent of the Sahara, are the areas where the sand moves, devoid of vegetation. On the other hand, the good water storage of stationary sands allows the growth of *Calligonum comosum*, Sahara broom (*Genista saharae*), and *Tamarix articulata*.

In the Saharan alpine mountains of Hoggar and Tibesti, Sudanese cassia and spider flowers can be found below 1,700 m (*Cleome* sp.). Between 1,800 and 2,400 m, a Mediterranean altitudinal zone extends with field wormwood (*Artemisia campestris*), endemic myrtles (*Myrtus nivellii*), olive tree (*Olea laperrini*), and pistachios (*Pistacia vera*). Higher than 2,500 m, it transitions into Artemisia communities (*A. herba-alba*) and Aristida grasslands.

Where open water or groundwater is present in the midst of the desert or semi-desert, oases form. The natural vegetation includes thorny shrubs as well as reed stands and trees: *Acacia* spp. and Doum palm (*Hyphaene thebaica*). Depending on the type of water supply, there are **river oases** with river flooding (Nile), **mountain foot oases** with groundwater outflow (Atlas Mountains, Tibesti Mountains), **spring oases**, which are often lined up in valley furrows, and **groundwater oases**, where water is available at a depth accessible to plants. Today, the cultivation of crops such as date palms, vegetables, fruit, and grain fields dominates in the oases. The most important tree of the North African oases is the date palm. It tolerates light frost but requires desert heat to ripen the fruits.

Under comparable moisture conditions of the semi-desert belt on the southern border of the Sahara, but with winter rainfall, the vegetation on the northern edge of the Sahara consists almost exclusively of tough tussock grasses and semi-shrubs. The tussock grass semi-deserts are mainly formed by esparto grass (*Stipa tenacissima*), which is harvested as a fiber material. On clay plains, semi-shrubs, particularly wormwood (*Artemisia* sp.), replace the tussock grasses. Tree and shrub growth is almost entirely absent in these formations, while further south in the desert, acacias still occur in the low lines of the wadis. The area of the tussock grass semi-deserts already belongs to the Mediterranean floral region. In the moister marginal parts, the grass growth becomes denser, and the semi-deserts transition into subtropical wintergreen dry steppes. Winter rainfall, ranging from 200 to 400 mm allows an evenly distributed cover of shrub and grass. The use of this natural steppe is currently mainly in rainfed agriculture.

## 8.4 The East and South African Highlands

Most East African highlands are occupied by the previously mentioned plant formations typical of the wet-dry and dry tropics. Near Lake Victoria, these are predominantly **wet savannas and semi-evergreen rain forests**. To the south, an almost continuous distribution extends beyond the southern tropic, comprising grasslands of the **dry savanna** interspersed with **Miombo forests**. In Kenya, **thorn savannas** predominate, which give way to thorny shrubs and succulent formations on the Somali peninsula, eventually transitioning northward into **dwarf shrub**

**semi-deserts**. The local diversity of vegetation is largely a result of the morphology. From the large zonal arrangement, the vegetation of the higher altitudes consists partly of upland evergreen forests, often in conjunction with grassland and heathland in a kind of mosaic. Examples are the **termite savannas** in the Iringa highlands in central Tanzania at an altitude of 1,500 m, which also occur at lower altitudes in the south in Mozambique, and with narrowly changing altitude levels on the high volcanoes such as Mount Kenya, Kilimanjaro, and the Virunga's. These rise from the grasslands through the stages of tropical mountain forests and cloud forests up to and beyond the tree line and have in the height above it the particularly pronounced Afromontane forests, the "ericaceous" belt, and the afroalpine high altitude communities of the **tropical high mountains**. They can also be found on the highest heights of the Ethiopian highlands, which, in its southern, more humid regions, and especially along its southwestern edge, have mountain forests of tropical character. On the inner high plateaus and in the northern mountain regions, high savannas and dry forests predominate. The northern slope of the mountains features thorn savannas, and the northern foothills, corresponding to the latitudinal location, are semi-deserts.

The transition from the submontane zone of the tropical rain forest to the montane rain forests occurs in the mountainous regions of East Africa, depending on exposure at 1,500–1,700 m. Due to the dense settlement in the area of the Central African Ridge and the volcanic massifs, the potential natural vegetation of both vegetation zones is largely degraded or completely destroyed and only preserved in the Ruwenzori and Virunga National Parks. Up to an altitude of 2,200–2,400 m, the **montane rainforest zone** (or mountain forest zone) follows with the evergreen conifer species *Podocarpus milanijanus* and the East African camphor wood (*Ocotea usumbarensis*). In the degraded areas up to the limit of permanent settlement at an elevation of 2,400–2,500 m, a so-called **tall grass savanna** with finger millets (*Digitaria* sp.) and/or acanthus bushland (*Acanthus pubescens*) is formed as secondary vegetation. From 2,300–2,800 m, up to the forest line at about 3300 m altitude or up to the tree line at around 3,800–4,000 m, the bamboo and **mountain bamboo zone** and the **Hagenia zone** follow (see Table 8.1 and Fig. 8.12). Above follows the montane rainforest zone of the ericaceous, dominated by giant heaths and heaths-like plants with mighty *Erica arborea* and Sphagnum-/Breutelia-moss areas.

At the large volcanic massifs, the **afro-alpine vegetation zone** begins with giant lobelias (*Lobelia* sp.) and giant groundsels (*Senecio* sp.) at 3,700–3,800 m altitude, which extends to the vegetation line at around 4,400 m. At an elevation of 4,600 m, the climatic snow line is finally reached. In the Highlands of Ethiopia, land use extends up to a height of 3,700 m (see Table 8.2).

On the eastern slope of the South African highlands, south of 24° S, there are coastal rain forests with moderate but year-round rainfall, which transition into subtropical rain forests in valley gorges and protected mountain locations. On the plateaus of the eastern highlands, on the inland side of the Great Escarpment from Lesotho to Transvaal, evergreen fellfields ("**Veld**") with low grass growth and precipitation in summer are widespread. With increasing summer dryness to the west, these transition into open **thornbush savanna**, to which, as already mentioned, further to the west, the closed belt of the **dwarf shrub semi-deserts** adjoins, which extends between the thorn savannas of the Kalahari and the Namib Desert northwards up to Southern Angola. In the 800–1,300 m elevated endorheic inner Kalahari Basin, the vegetation corresponds to the increasing rainfall from south to north: In the south, there is a **dwarf shrub-succulent formation** with dry grassland, and to the north **thorn-tree and desert grass** with Acacia savanna and transition to **thornbush savanna**. The **Namib** encompasses an approximately 100 km wide and 2,000 km long coastal strip from southern Angola to the Republic of South Africa. Due to the cold upwelling water of the Benguela Current off the coast, the Namib receives less than 20 mm

**Table 8.1** Altitude zones of vegetation on the Ruwenzori, DR Congo. Based on data of Wiese (1997)

| Elevation [m] | Mean annual precipitation [mm] | Mean annual temperature [°C] | Climate | Vertical vegetation zonation |
|---|---|---|---|---|
| Summit 5110<br>Snow line<br>approx. 4600 | | | Nival, snow zone<br>(cirque glacier) | |
| Vegetation line<br>approx. 4400 | 1600–1700 | 1.5 | Tierra helada<br>(soil flow,<br>patterned ground) | Alpine and subalpine zone,<br>Paramo vegetation or afro-alpine vegetation:<br>*Senecio* sp.; *Lobelia* sp.;<br>*Alchemilla- u. Helichrysum-*<br>Formation |
| 4000 | 1900 | 5 | | |
| 3700 / 4000<br>(Tree line<br>approx. 3700) | 1900 | 5 | | Ericaceous belt *Erica bequertii*, ubiquitous Usnea-beard lichens |
| (Forest line<br>approx. 3300) | | | Tierra fria<br>(rich in fog) | |
| 2700 / 2800 | 3000 | 12.4 | | |
| 2600 / 2700 | 3000 | 12.4 | | Mountain bamboo zone<br>*Arundinaria alpine;*<br>extended areas of<br>Sphagnum |
| | max. 3950 | | | |
| 2200 / 2300 | | 16 | | |
| 2000 | 2250 | 16 | Tierra templada,<br>excessive<br>precipitation | Mountain rain forest<br>*Albizzia gummifera;*<br>*Podocarpus*<br>milanjianus;<br>rich in liana plants |
| 1600 | 1800 | 20.5 | | |

of precipitation and is therefore almost devoid of vegetation in its coastal part between 18° and 28° S and in the southern **Sand-Namib**. Even the well-known characteristic plant of the Namib, the endemic *Welwitschia mirabilis,* is only found in a few places. The **Gravel-Namib** in the north is divided into the Outer Namib and the Inner Namib. While the Inner mainly has dry grasslands (*Stipagrostis* spp.), the outer features crust lichens, window algae, xerophytes, and small succulents (Aizoaceae, Asteraceae: Othonna, Kleinia) on rocky sites and inselbergs. At the border to the steeply descending Great Escarpment, among rock blocks, there are Cyphostemma, a plant genus within the family of grapevines, *Moringa ovalifolia* from the monogeneric family of the horseradish tree (*Moringacea*), and poikilohydric, so-called resurrection plants like *Myrothamnus flabellifolia*. Similarly, to the east, dwarf shrub semi-deserts also adjoin the desert to the south**.** For the Oranje—Olifant and small Namaqualandwith low winter precipitation, a species-rich succulent vegetation is as typical as for the "Succulent Karoo" of the southern Kalahari.

**Fig. 8.12** Tree with terminally tufted leaves (giant groundsel, Dendrosenecio kilimanjari, formerly *Senecio* spp.), also known as giant senecio, in Kilimanjaro National Park in Tanzania at an elevation of about 3,500–4,000 m in front of the summit of Kibo in the background. *Photo* S. Auth, Mauritius images

The **Karoo**, which encompasses South Africa north of the Great Escarpment and southern Namibia, like the southernmost part of the Kalahari, already belongs to the **Cape Floristic Region**. It is a vegetation-geographical hotspot and is characterized by succulent-rich semi-deserts with an extremely species-rich, endemic, and ecologically very specially adapted flora. The Great Karoo between the Cape Mountains and the Great Escarpment is today, due to overuse, especially by overgrazing, predominantly transformed into an Asteraceae dwarf shrub formation and a less dominant succulent vegetation. Block scree shrubs (*Cussonia spicata, Olea verrucosa*) and dry grasslands (Eragrostis, Sporobolus, Stipagrostis) characterize the Upper Karoo or the Central Plateau (occasionally also referred to as "Highveld"), which drops from the Great Escarpment from 1,800 to 1,000 m to the Orange River.

**Excursus: Madagascar**

The natural vegetation of Madagascar is determined by the mountainous character of the island and the influence of the Southeast Trades of the Southern Hemisphere. The eastern side of the island, which receives orographic rainfall, is home to rainforests and mountain forests. At the foot of the eastern slope, there are also semi-evergreen or rain-green rain forests. The largest part of the island's interior is dominated by tall grasslands. On the eastern part of the island, wet savannas and wet forests are found. In the uplands are high-altitude savannas, in places with remnants of evergreen sclerophyllous forest. To the west, dry savannas prevail instead of dry forests, which, however, are still preserved in a broad strip along the west coast. The driest part of Madagascar in the southwest of the island is characterized by thorn savanna and open succulent shrubs. The grasslands are fire-induced, species-poor, permanent communities. While the potential natural vegetation formations have 5,000 plant species, the grassland on an area seven times larger contains only 125 species. Only four of the 19 dominant grass species are native to Madagascar. The trees and shrubs that occur in the grasslands are based on a strict selection of fire-resistant species.

## 8.5 Mediterranean-Type Shrublands of North Africa and the Southwestern Cape

Then, the distinctive climate of regions with a warm, dry summer alternating with a cool, rainy period is characteristic of the Mediterranean regions of North Africa and the southwestern

**Table 8.2** Agroclimatic altitude zones and land uses in Ethiopia. Modified from Wiese (1997) and Hurni (1986)

| Elevation [m a.s.l.] | <900 | 900–1400 | >1400 |
|---|---|---|---|
| >3700 | | | **High Wurch**<br>A: Frost line<br>B: not existing<br>C: Black soils<br>D: Mountain grassland, no trees |
| 3700 – 3200 | | **Wurch (when humid)**<br>A: Oats, 1 harvest/year<br>B: Drainage rarely used<br>C: Black soils, degraded<br>D: Erica, Hypericum | **Wurch (when wet)**<br>A: Oats, 2 harvests/year<br>B: Drainage widely used<br>C: Black soils, severely degraded<br>D: Erica, Hypericum |
| 3200 – 2300 | | **Dega (when humid)**<br>A: Oats, wheat, teff, 1 harvest/year<br>B: Occational terracing<br>C: Brown soils with high clay content<br>D: Juniperus, Hagenia, Podocarpus | **Dega (when wet)**<br>A: Oats, wheat, barley 2 harvests/year<br>B: Widely used Drainage<br>C: Dark brown soils with high clay content<br>D: Juniperus, Hagenia, Podocarpus, Bambus |
| 2300 – 1500 | **Weyna Dega (when dry)**<br>A: Wheat, teff, rarely corn<br>B: Widely used terracing<br>C: Light brown to yellow soils<br>D: Acacia savanna | **Weyna Dega (when humid)**<br>A: Millet, corn, sorghum, wheat, barley, teff<br>B: Widely used terracing<br>C: Reddish-brown soils<br>D: Acacia savanna, Cordia, Ficus | **Weyna Dega (when wet)**<br>A: Corn, barley, teff<br>B: Widely used drainage<br>C: Red soils with high clay content, highly wheath-ered, gully erosion<br>D: Savanna with Ficus, Acordia, Acacia |
| 1500 – 500 | **Lower Kolla (when dry)**<br>A: Teff, rarely millet<br>B: Terraces for water storage<br>C: Yellow sandy soils<br>D: Shrub savanna domina-ted by Acacia | **Upper Kolla**<br>A: Millet, rarely Teff, cowpea<br>B: Widely used terracing<br>C: Yellow silty soils<br>D: Savanna with Accacia, Ficus, Erythrina, Cordia | |
| <500 | **Bereha**<br>A: Irrigation farming<br>B: not existing<br>C: Yellow sandy soils<br>D: Low tree and shrub savan-na dominated by Acacia | | |

Mean annual precipitation [mm]

A: Cereal and legume crops grow
B: Traditional soil conservation
C: Soil type
D: Natural vegetation/forest

corner of Southern Africa. The natural vegetation in the Maghreb region of North Africa is a direct reflection of the climatic conditions in its phytogeographical imprint, modified only by the altitude and the parent material. The essential factor of distribution and potential use is dryness. The natural vegetation can be divided into three zones: **Mediterranean, subalpine and alpine mountain forests, semi-shrub and steppe grass vegetation** (200–400 mm annual precipitation), and **diffuse, contracted vegetation of the semi-desert** with less than 200 mm annual precipitation. Especially in the Maghreb states, the increasing population and land use pressure through expansion of arable farming, livestock farming and logging leads to widespread degradation of the potential natural vegetation to a **bush forest (Macchia)**.

The extreme northwestern corner of the continent with the Atlas Mountains belongs with

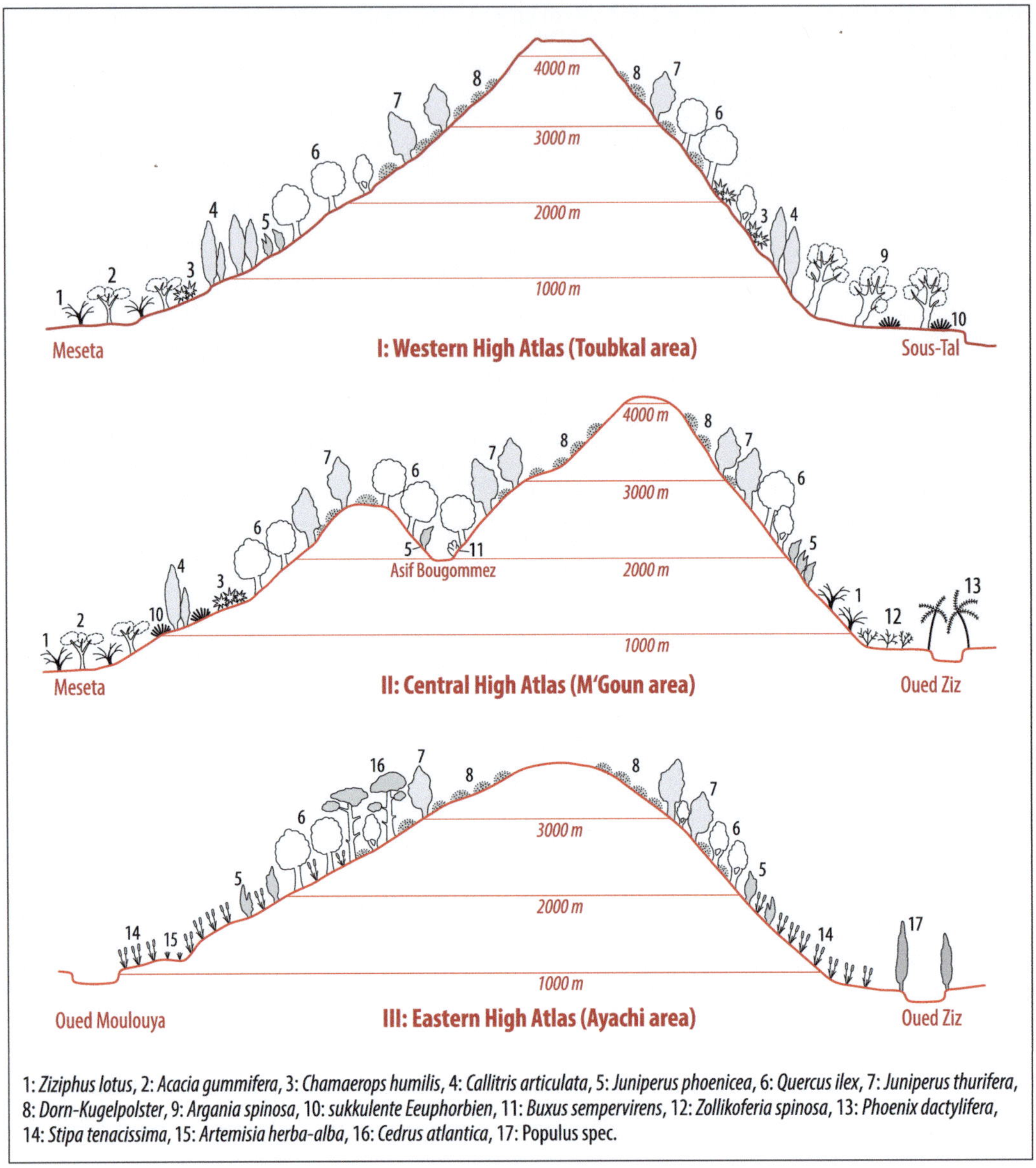


**Fig. 8.13** Zonation of vegetation across the High Atlas in SW Morocco east of Marrakech. The profiles show an increase in continentality from west (Jebel Toubkal) to east (Jebel Ayachi). *Source* Rauh (1952) in Müller-Hohenstein (1981)

its evergreen oak forests, Mediterranean shrubs of the Macchia type and coniferous dry forests completely to the **Mediterranean vegetation area.** In the Atlas Mountains, four altitudinal zones of vegetation can be distinguished. The cork oak zone is succeeded by the holm oak zone. Deciduous shrubs and mountain coniferous forests with the Atlas cedar follow in the cedar and fir zone, and Phoenician juniper represents the characteristic plant formation of the highest altitude zone of the subtropical alpine mountains (Fig. 8.13).

The extreme southwest corner of the continent shows the characteristic **Mediterranean hard-leaf vegetation with Macchia shrub formations**. On the Cape

Peninsula, it occurs at sea level, but not coastal; it only occurs at higher elevations. All formations of the **Cape flora** (the so-called **Fynbos**) are extremely rich in endemic species. For example, the family of Ericaceae is represented by around 400 species. The formation of evergreen forest, which grows up to 15 m high, only occurs in this winter rainfall area on the wettest slopes and the gorges of Table Mountain. Further east, on a narrow strip of the south coast, however, where precipitation falls all year long, an evergreen subtropical rainforest occurs rich in liana plants and some conifers (Podocarpus) as well.

## References

Bourgeron PS (1983) Spatial aspects of vegetation structure. In: Golley FB (ed) Tropical rain forest ecosystems. Ecosystems of the World 14A, Amsterdam

Clos-Arceduc F (1956) Etude sur photographies aériennes d'une formation végétale sahélienne: la brousse tigrée. In: Bulletin de l'IFAN. No spécial, Dakar

Deng Q, McMahon DE, Xiang Y, Yu C, Jackson RB, Hui D (2017) A global meta-analysis of soil phosphorus dynamics after afforestation. New Phytol 213(1):181–192

Hornetz B, Jätzold R, Litschko T, Opp D (1992) Beziehungen zwischen Klima, Wasserverhältnissen und Anbaumöglichkeiten in marginalen semiariden Tropen mit Beispielen aus Nord- und Ostkenya. Mat Z Ostafrikaforsch 9. Trier

Hurni H (1986) Guidelines for development agents in soil conservation in Ethiopia. Adis Abeba, p 123

ICO (2022) Erntemenge von Kaffee weltweit in den Jahren 2003 bis 2021 (in Millionen Säcken à 60 Kilogramm). Statista. Statista GmbH. Zugriff: 18. Februar 2023. de.statista.com. https://de.statista.com/statistik/daten/studie/12508/umfrage/weltweite-produktionsmenge-von-kaffee-seit-2003/

Menaut JC (1983) Analyse quantitative des ligneux dans une savane arbustive préforestière de Côte d'Ivoire. Géo-Eco-Trop 1:77–94

Müller-Hohenstein K (1981) Die Landschaftsgürtel der Erde. Teubner, Stuttgart, p 204

Rauh W (1952) Vegetationszonen im Hohen Atlas und dessen Vorland. Sitz ber Heidelb Akad Wiss 1

Schultz J (2002) Die Ökozonen der Erde. Ulmer, Stuttgart, p 332

Spinage CA (2012) African ecology—benchmarks and historical perspectives. Springer, Berlin Heidelberg, p 1562

van der Werf GR, Randerson JT, Giglio L, Collatz GJ, Mu P, Kasibhatla PS, Morton DC, DeFries RS, Jin Y, van Leeuwen TT (2010) Global fire emissions and the contribution of deforestation, savanna, forest, agricultural, and peat fires (1997–2009). Atmos Chem Phys 10:11707–11735

Walther H (1990) Vegetation und Klimazonen, 6th edn. UTB, Stuttgart, p 244

Wiese, B (1997) Afrika: Ressourcen, Wirtschaft, Entwicklung. Stuttgart Teubner, p 262

## Further Readings

Adams ME (1996) Savanna environments. In: Adams WM, Goudie AS, Orme AR (eds) The physical geography of Africa. Oxford University Press, New York, 196–210

Gießner K (1990) Geoecological control of fluvial morphodynamics in the Mediterranean subtropics. Geoökodynamik 11:17–42

Grainger A (1996) Forest environments. In: Adams WM, Goudie AS, Orme AR (eds) The physical geography of Africa. Oxford University Press, New York, pp 161–172

Herrmann SM, Anyamba A, Tucker CJ (2005) Recent trends in vegetation dynamics in the African Sahel and their relationship to climate. Glob Environ Chang 15:394–404

Hornetz B, Jätzold R (2003) Savannen-, Steppen- und Wüstenzonen. Braunschweig, Westermann, p 312

Keay RWJ (1959) Vegetation map of Africa South of the tropic of cancer. Oxford

Kull CA (2000) Deforestation, erosion, and fire: degradation myths in the environmental history of Madagascar. Environ Hist 6(4):421–450

Meadows ME (1996) Biogeography. In: Adams WM, Goudie AS, Orme AR (eds) The physical geography of Africa. Oxford University Press, New York, pp 161–172

Schulz E (1999) The southern margin of the Sahara in the Republic of Tchad. Vegetation, soil, and present pollen rain. Zbl Geol Paläont 1(5/6):483–496

# 9 Climatic Variations and Effects on Contemporary Climate Change

"**The Nairobi Declaration on Climatic Change**," adopted at the first international conference on global warming and its effects on Africa's climate, warned as early as 1990 that climate change would have significant ecological, economic, and political consequences. However, to this day (beginning of 2025), due to the size of the African continent, there is no universal trend regarding the development of climate conditions. There is, however, broad agreement that the climate change is primarily due to the accumulation of atmospheric gases such as $CO_2$ as a result of the combustion of fossil fuels and not due to natural climate change that were not $CO_2$-induced, such as at the end of the last glacial period (see Excursus: Sahara or Sahel? and Fig. 9.1). To this day, little research has been done, for example, on the influence of the three main climate drivers of Africa (**ITCZ, ENSO, West African Monsoon**) by climate change and their teleconnections. On the one hand, it is widely postulated that Africa will be disproportionately affected by climate change, with impacts expected to be more severe than in other regions of the world. On the other hand, there are few studies on the direct human influence, for example, whether the current dry period in the wet-dry tropics and dry subtropics of the continent is the result of climate change or caused by anthropogenic land degradation, possibly but also "only amplified". However, it is becoming increasingly clear today that the probability of occurrence and the magnitude of droughts, floods, and other extreme weather events will not only increase, but these events will also last longer each time. A catastrophic perspective, the dimensions of which become particularly visible in the case of potential droughts: Of the 54 states of the continent, 31 are directly affected.

**Excursus: Sahara or Sahel?—10,000 Years Between Pasture and Desert**

After extremely arid climatic conditions since the beginning of the last glacial maximum (LGM), which were characterized by the expansion of dunes located 400–600 km south of today's Sahara border, a humid phase started since the Late Glacial and especially since the Pleistocene/Holocene turn about 11,000 years ago, marked by lacustrine sediments. Lacustrine formations consisting of the siliceous shells of dead freshwater algae (diatoms) began between 10,000 and 9,500 years B.P. (Before Present). The lakes reached their greatest extent and maximum water depths of up to 30 m around 9,000 B.P. At this time, most of the southern central Sahara (Ténéré, Erg du Ténéré, Grand Erg de Bilma) was covered with lakes, which transitioned northward into a closed wetland (swamp zone). Lower lake levels occurred



R. Baumhauer, *Physical Geography of Africa*, Geography of the Physical Environment,
https://doi.org/10.1007/978-3-662-71406-5_9

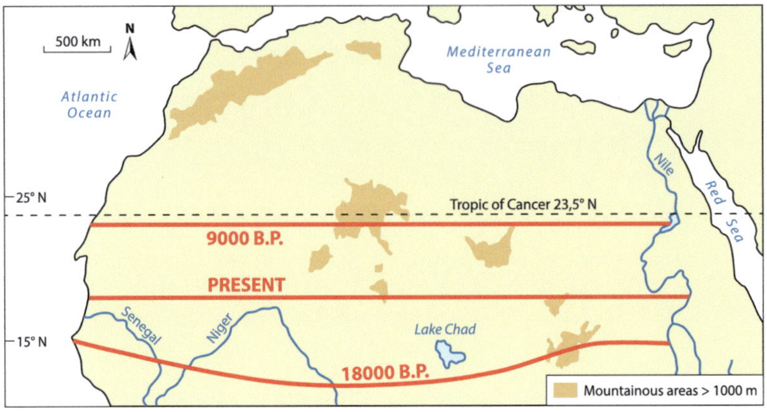


**Fig. 9.1** The changing position of the Sahara-Sahel limit

during several short dry phases (for example, during the "8.2 kiloyear event", the so-called "Misox fluctuation") until about 6,500–5,500 years ago. This was coupled with a profound change in palaeoecological conditions. Starting in the central Saharan plateau landscapes, the transformation of the extensive lake landscape into a wetland-dominated landscape occurred, which eventually turned into local swamp areas. They have given way to fully arid conditions throughout the area since at least 5,000 B.P. Pollen analytical studies have shown a change from a (Sahelian) savanna vegetation to more Saharan floral elements since 7,000 B.P. This aridification visible in the vegetation is at least partly due to the regular influence of humans in the contact area of the Sudanese to Saharan vegetation since the Middle Holocene (Schulz et al. 2009).

At the Pleistocene/Holocene turn in the Central Sahara, an annual rainfall rate of at least 400 mm is assumed. The rainfall was then tied to an interaction of the monsoon front with Atlantic-Mediterranean cyclones from the westerlies, significantly further south than today. During the Early Holocene and intensified in the Middle Holocene, the rainfall gradient from S (SW) to N (NE) increased, indicating a greater seasonality of rainfall, which hardly reached more than 200 mm per year. These fell predominantly as heavy rain and thus indicate a decreasing interlocking of the monsoon front and westerlies. Since 6,500 B.P. in the northern sections of the southern Central Sahara and from 5,000 B.P. in the southern parts, only the monsoonal rainfall regime is active, which was no longer sufficient for a widespread phreatic rise with corresponding groundwater recharge, for example, for the formation of permanent or periodic lakes. Interactions between westerlies and the monsoon front only occurred in exceptional cases, but when they did occur, they brought unusually high rainfall to the southern Central Sahara.

Studies in the Eastern and Western Sahara confirm the zonality of the palaeoclimatic conditions and suggest ecosystem structures during the Early Holocene moisture optimum in the Sahara, as they currently exist at about 13° N. This is coupled with a northward shift of the Sahel border by about 400–600 km.

It has been emphasized several times that, in contrast to the cool temperate climates, as we find them, for example, in Central Europe, there are no thermally influenced seasons like winter and summer in Africa, but the annual rainfall cycle dominates in the form of a typical alternation between rainy and dry seasons. Therefore, in Africa, changes in the hygric course of the year are more significant for climate fluctuations and climate change than temperature changes.

Comparative studies of the spatiotemporal patterns of precipitation for the decades 1931–1960 compared to 1961–1990 or 1902–2013 (see Fig. 9.2) show that during these periods, during the northern summer, precipitation in the Sahel decreased by more than 0.4 mm/day (in some cases by more than 30 per cent overall). On the Guinea coast of Ghana, Togo, Benin, and in parts of western equatorial Africa, on the other hand, it increased by around 10 per cent. This could be because in front of the **Dahomey Gap** (see Chap. 5), the sea surface temperatures tended to rise, thus weakening the upwelling of the cool equatorial current, which led to increased convection. The **IPCC (Intergovernmental Panel on Climate Change)** states in its 2022 report that from 1981 to 2010 in the Sahel, the frequency of extreme precipitation events has increased, and since the 1950s, the frequency of droughts has also increased (p. 1683).

The strong variability of precipitation in the Sahel region in the twentieth century becomes particularly apparent when viewed on a decadal basis. Recent climate models suggest that the above-average high rainfall in the 1950s and 1960s, as well as the Sahel droughts in the 1970s and 1980s, can be attributed to changes in the **Sea Surface Temperatures (SSTs)** in the tropics. Over the last 20 years, a positive rainfall trend has been recorded in the Sahel, likely due in part to the exceptionally dry conditions of the 1980s. Land–atmosphere feedbacks, changes in soil moisture, coverage and albedo, which together with climate or rainfall variability can further enhance positive and/or negative rainfall anomalies, are still not sufficiently taken into account in the climate changes in the Sahel (see Chap. 7 and Fig. 9.3).

According to the IPCC (2022), there is still no confirmed information available for Central Africa due to the scarcity of weather stations in the region. As a result, no reliable and empirically relevant data on rainfall patterns are currently available (p. 1684). For East Africa, the IPCC reports rainfall rates that have increased over large parts of East Africa since the 1960s in the months from October to December (p. 1685) and are projected to increase by up to 15 per cent. Currently, the largest percentage increases in rainfall are recorded in equatorial East Africa between September and November. This, along with temperature changes, will shift the current agro-ecological zones and influence agriculture in arid areas and irrigation farming in the future. When looking at crop yields under changed climatic conditions, it becomes clear that the harvest could decrease by up to 40 per cent for all examined crops, although the extent of the decrease depends on the respective crop and can, therefore, vary greatly. Thus, the IPCC report of 2022 (p. 1717) predicts a decrease in tea cultivation in Kenya of about 27 per cent compared to the year 2000, even under optimal conditions, with a warming of 1.8–1.9 °C. The cause of this is supposed to be the warming of the Indian Ocean, which leads to a stronger expression of the **Walker circulation** in this area (p. 1685). Since 1986 (until 2007), a decrease in rainfall has been recorded in East Africa during the rainy season between March and May, and since 2005, drought susceptibility has doubled (IPCC 2022, p. 1685).

In the southern summer, in the corresponding periods 1931–1960 versus 1961–1990, parts of Botswana and Zimbabwe experienced a 10 per cent decrease in rainfall, while rainfall increased by the same amount in the interior of Tanzania and northern Madagascar. In Namibia, Botswana, and southern Angola, the average annual rainfall activity statistically increased from 1980 to 2015, but this can be mainly attributed to a slight increase in extreme events. However, the number of drought years has not

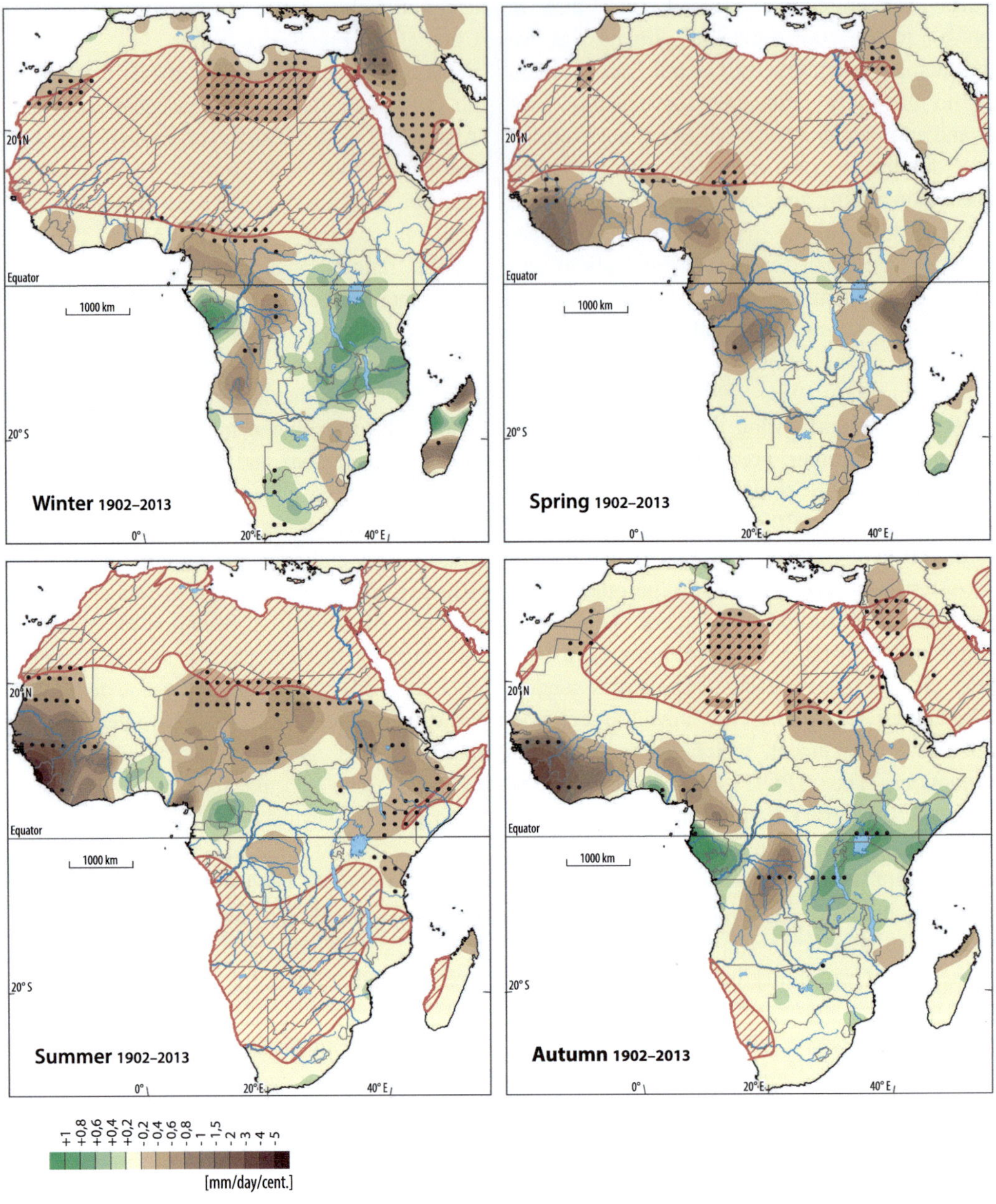


**Fig. 9.2** Change of mean seasonal rainfall between 1902 and 2013, expressed in mm/day$-^{1}$ century$^{-1}$. *Source* Thomas and Nigama (2018). The nonuniform contouring and shading interval are indicated via the color bar. Thick solid brown contours mark the 100 mm year climatological precipitation isoline, and brown hatching indicates regions where climatological precipitation 100 mm year—a precipitation threshold used for defining the Sahara Desert. Trends significant at the 95 per cent confidence level are denoted with black dots. Major rivers are shown in thin blue lines and country boundaries in thin gray lines

decreased. At the same time, declining rainfall rates were observed in the winter rainfall areas of South Africa (IPCC 2022, p. 1686). Overall, however, the changes in annual rainfall totals in the southern hemisphere are less significant than those north of the equator in Africa. Since rainfall activity in many parts of Africa has a strong seasonal component, it is particularly dependent

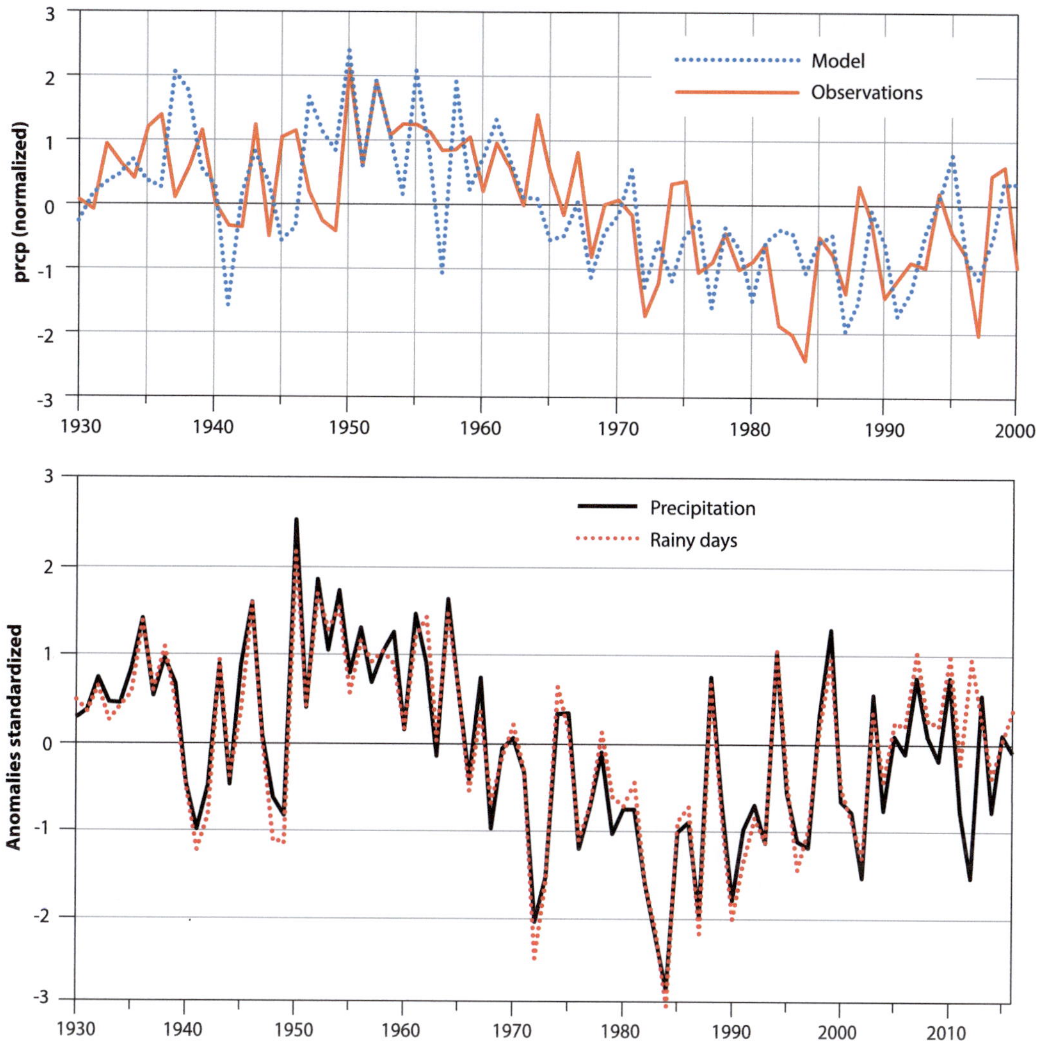


**Fig. 9.3** Homogenized time-series of annual Sahel rainfall anomaly. Observations used the average of stations between 10–20° N and 20–30° E (**b**) or 40° E, 35° E, model (**a**), each from July to September **a** from 1930 to 2000 according to Giannini et al. (2008) and **b** 1930–2018 according to Biasutti (2019). The correlation between observed and modeled indices of rainfall over 1930–2000 is 0.60 (time series are standardized to allow for an immediate comparison)

on the movements of the ITCZ. The change in seasonality with increased rainfall in the northern summer and less in the southern summer is most visible on the Guinea coast of West Africa. In the coastal regions of East Africa, however, rainfall seasonality has decreased, and in the Sahel of West Africa, it has remained constant.

In large parts of North Africa, the mean annual rainfall has already decreased from 1971 to 2000 (IPCC 2022, p. 1681). The number of rainy days with more than 10 mm of rainfall has decreased since the 1960s over the eastern regions of North Africa, and the number of dry days has increased. Contemporary climate models suggest that annual rainfall totals in many parts of Mediterranean Africa, the northern Sahara, and the southern winter rainfall areas will continue to decrease. Over the same period, extreme rainfall events have increased in the Maghreb (IPCC 2022, p. 1681, Fig. 9.4 and Chap. 5).

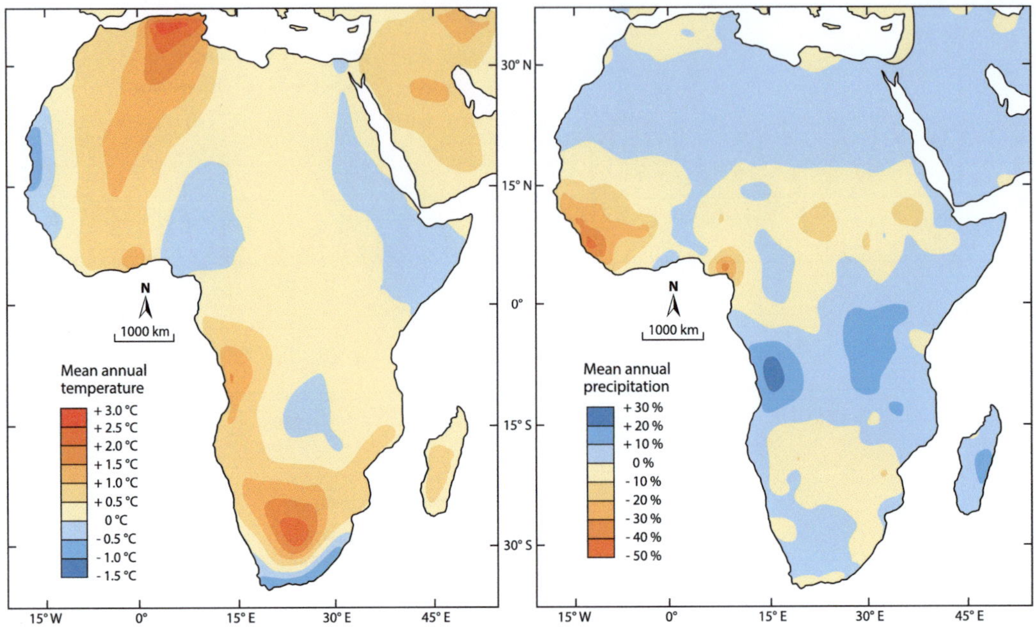


**Fig. 9.4** Changes in mean annual temperatures and mean annual precipitation rates between 1901 and 1995. *Source* Hulme et al. (2001)

Regarding interannual rainfall variability, the **ENSO phenomenon** appears to be one of the most important factors for some regions of Africa. This is at least correct for eastern equatorial Africa during the short rainy season (October and November) and southeast Africa during the main rainy season (November–February): In eastern equatorial Africa, high rainfall occurs during **El Niño**, in southern Africa, low rainfall under **La Niña** influence, especially in the months of the southern summer (Fig. 9.5). However, it is currently not possible to estimate how the development of the ENSO phenomenon and thus the variability of rainfall will evolve during climate change. The direct effects are likely to vary greatly spatially. While rainfall in East Africa is expected to increase, a drier and hotter climate is currently being modeled for South Africa, with a higher frequency of extreme events. It is also unclear how the cold-water upwelling system of the Benguela Current offshore will develop. If there is increased weakening here, resulting in more frequent southern advances of the warm Angola (Counter) Current offshore, this will not only have climatic effects on the lowland and escarpments of Angola, but also on the marine fauna. Fishing on the west coast of southern Africa is extremely dependent on the upwelling of cold, nutrient- and plankton-rich deep water.

**Excursus: ENSO Phenomenon**

The El Niño phenomenon refers to the periodic occurrence of a large, unusually warm water area in the eastern and central equatorial Pacific; the La Niña phenomenon refers to the opposite, i.e., an unusually cold-water area in the same part of the Pacific. The warming/cooling of the equatorial Pacific triggers the adjacent oceans and the atmosphere in different ways. Changes occur in atmospheric pressure patterns, the southern oscillation (SO), measured as the difference in standardized pressure values between the stations in Darwin and Tahiti, and represented as an index, Southern Oscillation Index

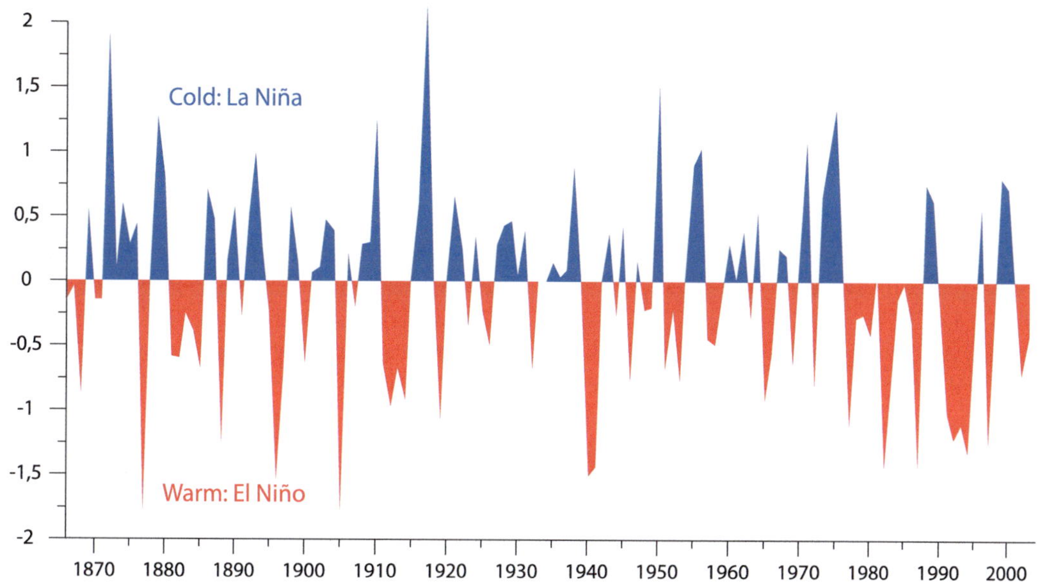


**Fig. 9.5** SOI (Southern oscillation index) anomaly between 1870 and 2000 from Rathmann (2009, p. 201)

(SOI). Due to the close relationship, these mechanisms are summarized as warm (El Niño) and cold (La Niña) ENSO phenomena. ENSO phenomena usually last three to six seasons (nine to 18 months), but sometimes also 24 months, and occur about every three to seven years, sometimes cold and warm events follow each other directly. El Niño conditions in the Pacific are associated with a warming of the western-central Indian Ocean and increased upper westerlies over Africa. Much of the ENSO influence on monsoon processes is driven by the SSTs of the Indian Ocean.

Although ENSO events have the strongest effect in the Pacific Ocean, an increased influence on the African continent has been observed in past ENSO events. Some pronounced droughts/floods, and precipitation anomalies are associated with it. In West Africa and especially in the Sahel of West Africa, the summer rains in the last 25 years are supposed to correspond with a high SOI. During warm ENSO events, anomalies in the SSTs are associated with a reduction in summer (July–September) rainfall and wetter conditions in June and October. A bipolar response to ENSO events was observed in eastern and southern Africa: in the east, in phase with warm ENSO events, and in the south, negatively correlated with these between October and December. In southern Africa, a strong relationship between warm ENSO events and droughts was also observed. Warm episodes that begin in April or May influence dry conditions in the following summer months (December to February, Fig. 9.5).

**Temperature changes** are also clearly detectable in Africa. While in equatorial Africa and parts of West and North Africa, temperature decreases of up to 0.5 °C can be observed, temperatures in the Sahel and large parts of southern Africa are increasing by up to 0.5 °C in the summer of the hemisphere. However, it should be noted that these changes only refer to mean temperatures and, therefore obscure the sometimes-contrasting trend of minimum and maximum values (especially at night vs. during the day). The IPCC describes in its fifth climate status report (2014) an increase in the global average temperature of 0.6 ± 0.2 °C for the twentieth century, with the intensity of warming varying regionally. The sixth assessment report states a change in the mean global surface temperature of 1.09 [0.95–1.2] °C (p. 10) for the period 2011–2020 in relation to the period 1850–1900. By the year 2100, a further temperature increase of 1.4–4.4 °C is predicted under the assumption of various scenarios. The warming is taking place faster in the land masses than over the oceans and is, therefore, likely to exceed the global average. The subtropical dry zones are expected to be more affected by the warming trend than, for example, the tropical rainforests (Fig. 9.6).

Other models simulate temperature changes of +1.5 to +2.5 °C for tropical-subtropical Africa north of the equator by 2025 (!) and assume that heat waves and dry periods will become more frequent as a result of a general temperature increase (among others, Paxian et al. 2016; Paeth et al. 2017). The increasingly dry periods play a key role in desertification processes, as they increase the vulnerability of soils to degradation in the affected areas and quantitatively and qualitatively affect the available water resources. In addition to crop failures and the endangerment of water supply, a further reduction of the already low soil moisture is expected in regions already threatened by desertification today.

Another aspect of climate change is the intensification of the hydrological cycle. Due to the general temperature increase, not only will evapotranspiration increase, but the atmosphere's capacity to absorb water vapor will also increase. This results in an increasing risk of heavy rainfall events, but also a further amplification of the greenhouse effect due to the increased water vapor content. Except for parts of the Mediterranean, however, the mid and high latitudes will likely primarily be by an increase in heavy rainfall events. On the other hand, a current analysis of about 30 climate models shows that there could also be increasingly heavy regional rainfall in the Sahel of Africa. The cause of this is a self-reinforcing mechanism that could start beyond a global warming of 1.5–2 °C. Several models indicate that the Sahelian climate changes abruptly when the temperature of the sea surface in the tropical Atlantic and the Mediterranean rises above a certain value, which represents a kind of tipping point. Precipitation then suddenly and strongly increases, coupled with an increase in the intensity of the southwest monsoon, which also extends considerably further north. More intense precipitation leads to increased surface runoff without a significant contribution to increasing the soil moisture storage. Rather, soil erosion is accelerated, which plays a significant role in degradation and desertification processes. In this context, the ambivalence of water in arid areas is once again emphasized. On the one hand, it is only available in limited quantities most of the time; on the other hand, it often shows its destructive power in the form of heavy rainfall events with soil erosion and flooding. However,

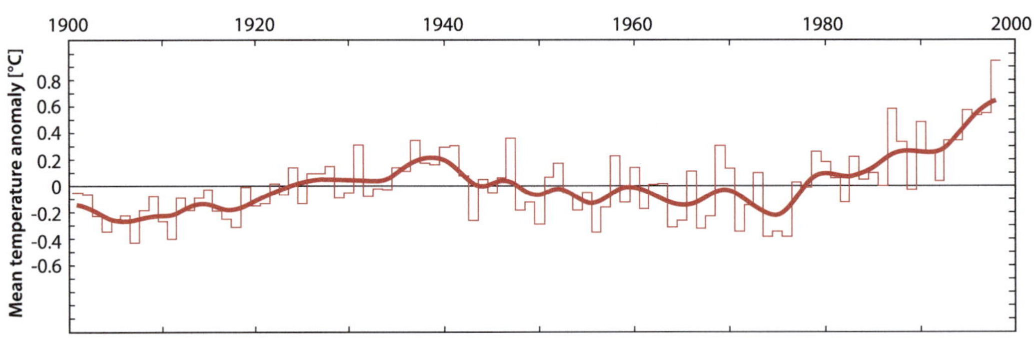


**Fig. 9.6** African annual mean temperature anomalies between 1900 and 2000. From Hulme et al. (2001)

the findings and forecasts regarding the changes in the hygric climate components precipitation and evaporation caused by climate change are still considered less certain than the temperature development.

In contrast, there is broad agreement that global warming has a major influence on the development of the water balance and its subcomponents on a global and regional scale. As a result, serious consequences are to be expected in terms of the availability of water resources. It has already been mentioned that the temporal and spatial distribution of precipitation is likely to change due to the general intensification of the hydrological cycle, driven by rising evaporation rates, in the form of an increase in the high and middle latitudes and a decrease in some areas closer to the equator. This, of course, also has consequences for the forecast of surface runoff. The IPCC (2014 and most recently 2022) considers an increase in the higher latitudes and Southeast Asia likely, while a decrease in runoff volumes is expected for large parts of the arid regions. A reduction in runoff caused by declining precipitation amounts with increasing evaporation is likely to intensify the general water scarcity and vulnerability to desertification, although the vulnerability to erosion could decrease at the same time. In addition, qualitative problems in water supply may arise. As evaporation increases with general warming, the change in surface runoff can differ significantly from that of precipitation. According to Maynard and Royer (2004, among others), a warming of 1–2 °C in combination with a 10 per cent decrease in precipitation reduces surface runoff by 40–70 per cent. Recent studies show that runoff volumes in the West African Sahel are increasing despite declining precipitation amounts. The so-called "**Sahel Paradox**" results from the combined effects of anthropogenic crust formation and general droughts, leading to reduced water absorption and storage capacity in the dried-out and often hardened soil. As a result, the precipitation cannot be absorbed by the soil and flows off the surface. In addition, increased evapotranspiration rates also affect the infiltration, the soil moisture, and thus the amount of plant-available soil water. In general, it can be stated that the water balance of arid and semi-arid regions is highly sensitive to climate variations. As a result, the risk of accelerated physical desertification processes cannot be excluded and is likely to be further intensified by the potential intensification of dry periods mentioned.

Examining the impacts of climate change on vegetation reveals several key aspects of interest. In addition to consequences for the metabolism of individual plants, there are implications for the species composition of plant communities and their distribution, as well as for agriculture. Rising temperatures increase the water demand of plants and influence the agroecological zones: High-altitude regions become more agriculturally usable year-round due to higher temperatures and lower frost risk. In the lowlands, plant growth is stimulated by higher temperatures, and the growth period is correspondingly shortened, which can affect the quality of the harvest. Higher temperatures will also directly affect yields by shortening the length of the growing season in regions where agriculture can be practiced at all, thereby reducing the yield potential. Temperature increases are particularly severe for pastoral societies, as not only perennial surface or well water must be available for livestock farming, but the quality of the pasture can change quickly due to the invasion of annuals and hard grasses and the encroachment of deep-rooted shrubs into grasslands (Kempf 2000).

Since the global distribution of vegetation (zones) is largely driven by climate in addition to soils, it is logical to conclude that climate change will also be reflected in the vegetation pattern. The plant world can react to changed climate conditions with succession. Conversely, the climate is influenced by changes in plant physiological processes, structural changes, and the coupling of vegetation to the carbon cycle. Overall, changes in the vegetation pattern caused by climate change are very likely at unchanged emission rates of greenhouse gases. When estimating the effects of climate change on agricultural yields, the spatially differentiated development of temperature and water availability is primarily relevant. Thus, temperature changes in crops that currently find their optimal temperature spectrum in a certain cultivation area will result in yield declines as soon as the thermal optimum range is left. The question of water availability is particularly important for the dry areas of the earth, as in these regions the hygric pattern usually limits plant production. In addition, other influencing factors such as $CO_2$-fertilization effects, plant diseases, changes in the vegetation period, etc., must be considered. The interaction of the mentioned factors and their value will ultimately decide regionally about agricultural productivity. While yield increases are considered possible for the higher latitudes, a negative development of agricultural yields with an increased risk of famines is likely for the arid and semi-arid areas. The increase in agriculturally usable land in the higher latitudes will, therefore, probably be accompanied by losses in the subtropics in the form of an expansion of steppes and deserts. In this case, the (land) usage pressure in the already desertification-endangered areas would increase once again.

In summary, stronger direct effects of climate change are already evident in Africa today. Due to its heavy reliance on agriculture and limited ability to adapt, the economy in Africa is overall more dependent on environmental conditions than in other parts of the world, further intensified by a rapid population growth, projected to increase from 1.34 billion inhabitants in 2021 (17.2 per cent of the world population) to around 4.3 billion in 2100, accounting for nearly 40 per cent of the world population (Table 9.1).

**Table 9.1** Effects of climate change on agriculture and water resources

| Climate change | Effect/impact | Consequence |
|---|---|---|
| $CO_2$ enrichment | Increased photosynthesis; reduced transpiration | Increased water use efficiency |
| Increased temperature | Faster plant growth, increased transpiration, but over a shorter growth season; reduced runoff and reduced groundwater recharge; higher demand for water for irrigation, bathing and cooling | Reduced economic agricultural yields; changes in water yields; higher stress and peak loads for water delivery systems |
| Sea level rise | Land loss; salt intrusion into coastal aquifers; movement of salt front estuaries affecting freshwater abstraction points | Reduction of coastal agriculture; reduced water quality in coastal areas; reduced groundwater abstraction |
| Change in seasonal precipitation | Change in soil moisture; change in runoff and groundwater recharge | Changes in agricultural yields; changes in projected yields of reservoir systems; changes in water quality |
| Change in spatial patterns of temperature and precipitation | Shift in agroclimatic suitability; shift in basin hydrology (surplus and deficit regions) | Changes in cropping systems; changes in infrastructure to supply water |
| Change in variability of precipitation (daily and inter-annual) | Change in water stress or between rainfall events; changes in peak runoff | Increased requirement for storage in marketed produce, subsistence crops, and water supply systems |
| Changes in drought hazard | Change in seasonal water stress or off-season water replenishment | Altered risk for agriculture production and water resources |
| Change in flood hazards | Change in risk in floodplain; change in area affected | Altered risk, especially for water resources; change in reservoir operations |

## References


Adhikari U, Nejadhashemi AP, Woznick SA (2015) Climate change and eastern Africa: a review of impact on major crops. Food Energy Secur 4(2):110–132

Biasutti M (2019) Rainfall trends in the African Sahel: characteristics, processes, and causes. WIRES Clim chang 10(4):22

Brooks N (2004) Drought in the African Sahel: long term perspectives and future prospects. Tyndall Centre Working Paper No. 61, p 31

Collier P, Conway G, Venables T (2008) Climate change and Africa. Oxf Rev Econ Policy 24(2):337–353

Downing TE, Ringius L, Hulme M, Waughray D (1997) Adapting to climate change in Africa. Mitig Adapt Strat Glob Chang 2:19–44

Flohn H (1965) ITCZ and meteorological equator. Studies on the meteorology of tropical Africa. Bonner Meteor Abh 5:49–57

Giannini A, Biasutti M, Verstraete MM (2008) A climate model-based review of drought in the Sahel: desertification, the re-greening and climate change. Glob Planet Chang 64:119–128

Hulme M (1992) Rainfall changes in Africa: 1931–1960 to 1961–1990. Int J Climatol 12:685–699

Hulme M (1996) Climate change within the period of meteorological records. In: Adams WM, Goudie AS, Orme AR (eds) The physical geography of Africa. Oxford University Press, New York, pp 88–101

Hulme M, Doherty R, Ngara T, New M, Lister D (2001) African climate change: 1999–2100. Clim Res 17: 145–168

IPCC (2014) Climate change 2014: synthesis report. In: Core Writing Team, Pachauri RK, Meyer LA (eds) Contribution of working groups I, II and III to the fifth assessment report of the intergovernmental panel on climate change. IPCC, Geneva, Switzerland, p 151

IPCC (2019) Summary for policymakers. In: Climate change and land, an IPCC special report on climate change, desertification, land degradation, sustainable land management, food security and green house gas fluxes in terrestrial ecosystems

IPCC (2022) Summary for policymakers. In: Climate change and land, an IPCC special report on climate change, desertification, land degradation, sustainable land management, food security and green house gas fluxes in terrestrial ecosystems

Jury MR, Enfield DB, Mélice J-L (2002) Tropical monsoons around Africa: stability of El Niño-Southern oscilllation associations and links with continental climate. J Geophys Res 107(C10): 15-1–15-17

Kempf J (2000) Klimageomorphologische Studien in Zentral-Namibia. Ein Beitrag zur Morpho-,Pedo- und Ökogenese. Diss. Fak. Geowiss. Univ.Würzburg, p. 613

Knox J, Hess T, Daccache A, Wheeler T (2012) Climate change impacts on crop productivity in Africa and South Asia. Environ Res Lett 7(3)

Mason SJ, Jury MR (1997) Climatic variability and change over southern Africa: a reflection on underlying processes. Prog Phys Geogr 21(1):23–50

Maynard K, Royer J-F (2004) Effects of "realistic" land-cover change on a greenhouse-warmed African climate. Clim Dyn 22:343–358

Nicholson SE (2000) The nature of rainfall variability over Africa on time scales of decades to millennia. Glob Planet Chang 26:137–158

Nicholson SE, Kim J (1997) The relationship of the El Niño-Southern oscillation to African rainfall. Int J Climatol 17:117–135

Nicholson SE, Selato JC (2000) The influence of La Nina on African rainfall. Int J Climatol 20:1761–1776

Obasi GOP (2005) The impacts of ENSO in Africa. In: Sum Low P (Hrsg) Climate change and Africa. Cambridge University Press, Cambridge, pp 218–230

Paeth H (2010) Der Klimawandel in Afrika: Physisch-geographische Befunde und Klimamodellsimulationen. In: Glaser R, Kremb K, Drescher A (eds) Afrika. WBG, Darmstadt, pp 22–31

Paeth H, Paxian A, Sein D, Jacob D, Panitz H-J, Warscher M, Fink A, Kunstmann H, Breil M, Engel T, Krause A, Toedter, Ahrens JB (2017) Decadal and multi-year predictability of the West African monsoon and the role of dynamical downscaling. Meteorol Ztschr 26(4):363–377

Paxian A, Sein D, Panitz H-J, Warscher M, Breil M, Engel T, Tödter J, Krause A, Cabos Narvaez WD, Fink AH, Ahrens B, Kunstmann H, Jacob D, Paeth H (2016) Bias reduction in decadal predictions of West African monsoon rainfall using regional climate models. J Geophys Res 121:1715–1735

Rathmann J (2009) Klima- und Zirkulationsvariabilität im südhemisphärischen Afrika seit Beginn des 20. Jahrhunderts. Dissertation, Universität Augsburg

Schulz E, Abichou A, Adamou A, Ballouche A, Ousseini I (2009) The desert in the Sahara. Transitions and boundaries. Palaeoecol Afr 29:63–90

Thomas N, Nigama S (2018) Twentieth-century climate change over Africa: seasonal hydroclimate trends and Sahara desert expansion. J Clim 31:3349–3370

# Index




R. Baumhauer, *Physical Geography of Africa*, Geography of the Physical Environment, https://doi.org/10.1007/978-3-662-71406-5

MIX
Papier aus verantwortungsvollen Quellen
Paper from responsible sources
FSC® C105338

If you have any concerns about our products,
you can contact us on
**ProductSafety@springernature.com**

In case Publisher is established outside the EU,
the EU authorized representative is:
**Springer Nature Customer Service Center GmbH**
**Europaplatz 3, 69115 Heidelberg, Germany**

Printed by Libri Plureos GmbH
in Hamburg, Germany